Everyone Flows

Everyone Flows

A Process Philosophy of Human Life

JOHN DUPRÉ

OXFORD
UNIVERSITY PRESS

Great Clarendon Street, Oxford, OX2 6DP,
United Kingdom

Oxford University Press is a department of the University of Oxford.
It furthers the University's objective of excellence in research, scholarship,
and education by publishing worldwide. Oxford is a registered trade mark of
Oxford University Press in the UK and in certain other countries.

© John Dupré 2025

The moral rights of the author have been asserted

Published in the United States of America by Oxford University Press
198 Madison Avenue, New York, NY 10016, United States of America.

British Library Cataloguing in Publication Data
Data available

Library of Congress Control Number: 2024947800

ISBN 9780198966647 (pbk.)
ISBN 9780198941828 (hbk.)

DOI: 10.1093/9780198941866.001.0001

Printed and bound by
CPI Group (UK) Ltd., Croydon, CR0 4YY

The manufacturer's authorised representative in the EU for product safety is
Oxford University Press España S.A. of El Parque Empresarial San Fernando de Henares, Avenida
de Castilla, 2 – 28830 Madrid (www.oup.es/en or
product.safety@oup.com). OUP España S.A. also acts as importer into Spain
of products made by the manufacturer.

For Jackson Theodore Dupré

Contents

Introduction

Like many books of philosophy this one has a long history. The necessity of a process philosophy of biology gradually dawned on me during the time I was directing what was then the Economic and Social Research Council (ESRC) Centre for Genomics in Society (Egenis), as I began to appreciate more fully the problems with seeing the genome as a more or less fixed thing composed of distinguishable parts, genes. I believe that the necessity of a process philosophy first became clear in my mind during the time when I had the privilege of spending a term in Amsterdam in 2006, as the Spinoza Professor of Philosophy. Indeed, though I think such conversion stories should be treated with some scepticism, I do seem to remember exactly such a Damascene moment.

I recall taking an exploratory walk around the area of Amsterdam where I was staying and wandering unexpectedly into the famous Albert Cuyp market. On my way there, I had been wondering about the increasing pedestrian density, and indeed I found my way to the market by following the gradient of this growing flow. As I reflected on the hive of human interaction that constituted the market, it struck me that it would be wholly wrong to think of this as some kind of thing, constituted by a set of component parts—stalls, people, commodities, and so on. Rather, the market was constituted by a constant flow of people bringing and buying goods, setting up stalls, and so on. The market had a diurnal cycle: every morning stallholders arrived, busily established their places of work, and waited for the flow of people to arrive and transact business with them. At night, the goods were taken away and the market entered a near dormant state, populated by just the occasional passer-by, policeman, or security person.

The market, then, was a process, maintained by this flow of people and stuff. If the flow stopped the market would cease to exist. It immediately struck me that much the same should be said about an organism, or for that matter a genome. These were not fixed structures, contingently open to accidental interventions, but processes, whose persistence from moment to moment depended on activity. It was not news that the persistence of an organism depended on trillions of microscopic chemical and physical events every second for it to continue to exist. But I don't think I had fully appreciated until about then the distance of this form of existence from that implied by our dominant philosophical account of the world as composed of things and their sometimes, but not necessarily, changing properties. My first attempt to formulate this process-centred picture, at any rate, was in the Spinoza lectures that I delivered in Amsterdam, later published as *The Constituents of Life* (Dupré 2008).

Elaborating a processual view of life, the universe, and everything, and exploring some of its applications and implications, has been the main focus of my philosophical work for the subsequent fifteen years. There are various strands to this research. One strand is direct engagement with the philosophical tradition in metaphysics, in which the dominant view remains that the fundamental description of the world should be in terms of things, or more technically, *substances*,[1] and their properties. Some philosophers claim that there is something incoherent in the very idea of a world of pure process.

Of course, I do not believe this negative conclusion, and I have occasionally attempted to engage with the arguments that claim to establish it. The problem is that this disagreement tends to reflect another, if not deeper, at least prior difference in the view of how the relevant branch of philosophy, metaphysics, works. Metaphysics aims to make very general, even abstract, statements about what there is in the world and how the world works, and to this extent I consider myself to be a metaphysician and this book to be an exercise in metaphysics. However, the recent tradition in at least anglophone metaphysics has generally sought to ground its findings in a mixture of common sense (or 'intuition', what seems self-evidently true) and logic. It is considered prior to empirical science[2] and thus to provide constraints on what science can possibly tell us. If metaphysics tells us that a world of process is impossible, then we cannot interpret science as revealing to us a world of process.

I align myself rather with what is often described as naturalistic metaphysics, a metaphysics that is deeply entwined with the progress of empirical science. This does not mean merely a glorified science journalism, a wide-ranging description of scientific beliefs. It is, rather, like any metaphysics, an

---

[1] Substance is a concept first articulated in detail by Aristotle (it is a translation of the Greek word οὐσία, meaning being). It has a rich and varied history in the Western philosophical tradition, attempting to trace which would be far beyond the scope of this book. Here, I shall generally use the more familiar word 'thing', which with a bit of refinement in the next chapter will adequately serve my purpose and avoid some philosophical technicalities.

[2] Timothy Williamson (2013), one of the most admired of contemporary metaphysicians, takes logic to be a science. It is, at any rate, surely not an empirical science, which is my concern in this book.

attempt to provide a very general description of the world, but one in constant dialogue with, and answerable too, our best empirical science. As such, it also has the capacity to be critical of science, where this appears to be inconsistent with our best metaphysics. At many points in this book, I shall suggest that our thinking on many issues, scientific and philosophical, suffers from our commitment to a traditional thing/property metaphysics and would be much improved by the adoption of a processual perspective.

It is not, however, the aim of this book to engage very much with the arguments for and against the traditional metaphysics. I shall explain, over the first three chapters, why I think a processual metaphysics is so well suited to understanding central parts of contemporary biology. In the following three chapters I shall argue that this metaphysics offers major insights into fundamental philosophical questions about the human; whatever else they may be, humans are biological organisms. The overall strategy of the book can perhaps best be seen as abductive, an inference to the best explanation. If, as I believe, a process metaphysics is so helpful in illuminating issues in both science and philosophy, this is a very good reason for thinking that it is true.

To summarize the contents of the book in slightly more detail, in chapter 1 I sketch the deepest divides between a process- and a thing-centred metaphysics, and show (very briefly) how the former provides a satisfying account of our overall picture of the history of our universe. I then outline several reasons why we should think of the biological organism as a kind of process rather than a kind of thing. In chapter 2, I explain the concept of a lineage, which I take to be a kind of process central to the proper understanding of evolution. This chapter is the most scientifically technical in the book, and the less scientifically minded reader might wish to skim over some parts of it. The general message will have some important implications for some of what follows, however. In chapter 3 I describe some of the vital and complex relations between organisms that are generally considered under the rubric of symbiosis. I take up in much greater detail an argument from chapter 1, that the profound nature of these interconnections provides a serious obstacle to thinking of the living world as divided into discrete entities in the way supposed by thing metaphysics.

In chapter 4 I turn explicitly to the human and the ancient philosophical question about personal identity, what it is for a person to persist through time. As should have become clear in chapter 3, a fundamental distinction between the two metaphysical systems I am considering is that a process metaphysics does not assume that either the spatial or the temporal boundaries of an entity are clearly fixed. Many such boundaries are conceptual, imposed by us for a variety of pragmatic reasons. This allows us a much more relaxed view of a range of questions about the beginning and end of a human animal or a human person and the temporal boundaries we draw within the life cycle of the human. In chapter 5, I consider questions first of human nature, then of subkinds within the human species. A generally sceptical attitude to the former leads to an equally critical attitude to the latter. The various divisions we make within the human species, sex and gender, race, and many more, are not to be seen as distinguishing different kinds of thing with different essential properties, but as pragmatic classifications of different developmental trajectories within a massively plastic species (or lineage). Finally, in chapter 6, I address the question of agency and free will. Here again, process philosophy provides a radically different perspective. The philosophical tradition has generally seen the world as deterministic, governed by strict laws that determine everything that happens, a view that has famously led to worries about freedom of the will. A process world, as I see it, is one in which the order assumed by determinists is a rarity acquired only by complex eddies within the chaotic world of process, and never fully. Humans are to be understood not as anomalous exceptions to the ordered world, as libertarian philosophers have often argued, but rather as exceptional islands of order within a largely chaotic and disordered world. Our sense of freedom, I argue, can be located readily within such a picture.

One final note. My training and expertise are entirely in the Western philosophical tradition. Both Indian and Chinese philosophy have, I am reliably told, paid more attention to process thinking than has Western philosophy. But I must leave it to others more knowledgeable to explore the connections between the ideas developed here and related thinking in these traditions.

1
Why Life is All Process

A World of Things

Let me begin by locating this book—and myself—on an intellectual map. My main disciplinary affiliations are with the philosophy of science, especially of biology, and metaphysics. Unlike some of my colleagues in both these fields, I take them to be closely intertwined. Metaphysics is notoriously hard to define, and its definition is controversial, but by it I mean something like the attempt to provide a general and wide-ranging understanding of reality. In the oft-quoted words of the twentieth-century philosopher of science Wilfred Sellars (1962), we are concerned with 'how things in the broadest possible sense of the term hang together in the broadest possible sense of the term'. Even more controversial than the definition of metaphysics is its correct methodology. In opposition to what is still the majority view, that metaphysics is an a priori discipline, I take it to be ultimately empirical, aiming at contingent truths about the world that draw their support from experience, generally as filtered by the methods of science. But unlike some of my colleagues in the philosophy of science, I do believe that there is an important and distinctive field of metaphysics, one that draws on the findings of science, but aims at truths at a more abstract level than can be accessible to the experimental methods of science. This is a project sometimes known as naturalistic, or scientific, metaphysics.

Underlying my own philosophy is a more specific broad metaphysical thesis, the implications of which will be the main topic of this book, that we live not in a world of things but in a world of process. This is a debate that can be traced throughout the history of philosophy at least to the pre-Socratics. Two of the best-known pre-Socratic philosophers, working in the late sixth and early fifth century BCE, were Parmenides and Heraclitus, famous, respectively, for asserting that nothing changes and that everything changes. Parmenides' doctrine is obscure, partly because his views are deduced entirely from the 205 lines or so that survive of an opaque metaphysical poem,

Everyone Flows. John Dupre, Oxford University Press. © John Dupré 2025.
DOI: 10.1093/9780198941866.003.0001

and partly because it seems so obviously to violate common sense: we observe change every moment.

However, Parmenides was taken by many of his contemporaries to have provided powerful arguments that change was impossible. Leucippus and his better-known student Democritus introduced an interpretation of the Parmenidean doctrine that remains influential to this day. What really exist are the eternal unchanging elements of reality, atoms. The passing show that we experience as a changing world is merely the rearrangement of atoms into different relations to one another.

This kind of atomism was a central philosophical element of the scientific revolution of the sixteenth and seventeenth centuries, and to this day is often considered a central feature of a scientific world view. It motivates the continuing ideal of reductionism, which sees the ultimate goal of all sciences of complex entities, from chemistry to sociology, as accounting for the behaviour of their subject matters in terms of the particles of which they are composed and the laws of physics. This is summed up by Robert Boyle (1627–1691), a leading figure in the scientific revolution: 'the Naturalist ... in explicating particular phenomena, considers only the Size, Shape, Motion (or want of it), Texture, and the resulting Qualities and Attributes of the small particles of Matter' (Boyle 1666). This widely shared view amounted to a decisive victory for the world view of Parmenides and Democritus.

The view that I shall defend and explore in this book is that this atomistic reductionism is entirely misguided. But not, as is accepted by a growing number of philosophers and even scientists, because the world is just too complex for the reductionism it proposes ever to be feasible, but because the whole atomistic vision is mistaken. The neo-Parmenidean world is a world of things: eternal things, atoms, and more or less stable things that are structures of atoms. I advocate a return to the world view of Heraclitus: there are no stable things at all. Everything always changes.

Let me make a brief nod here to my own philosophical history. In 1993 I published a book, *The Disorder of Things: Metaphysical Foundations of the Disunity of Science*, arguing against the still standard metaphysics of an ordered, deterministic world in which science was centred on the discovery of the essences of things and the explanation of their behaviour by deduction from the properties of their microscopic constituents. It was more than a decade later that I realized that what I was really groping to understand was a world of process. The rejection of the orderly, mechanistic world bequeathed us by the scientific revolution is necessary background for the processual world I now defend. I shall draw on that earlier work from time to time in this book. But now I shall try to explain what is at stake in the difference between a world of things, such as the atomistic world of Democritus or Boyle, and the world of process that I shall defend.[1]

A World of Process

The most fundamental difference between things and processes is that things are stable by default; not necessarily forever, unless they are atoms, but for some period of time. When they change, or cease to exist, we look for some explanation. We often think of causality, a fundamental element of most metaphysical systems, as the domain of explanations of the changes that things undergo. For processes, by contrast, stability is what primarily requires explanation. Most processes have no obvious stability; in the famous words attributed to Heraclitus, 'Everything changes.' But some entities in the world do appear stable, and indeed it is the stability of these entities that makes it natural for us to think of the world as composed of things. The challenge for the Heraclitean is to explain how these apparent things are in fact just temporarily stabilized processes. In fact, it turns out, such thing-like processes persist only for so long as further processes, internal or external, sustain and stabilize them. A paradigm of a simple persistent process is an eddy in a river. An eddy may maintain a more or less stable structure for a long period of time; but only for as long as the river flows past and through it. This, I shall suggest, is typical of many less obviously dynamic processes.

A thing, I have said, is stable by default, but there is more than that to the concept. Unfortunately, there is no clear and complete orthodoxy about the nature of things. The closely related concept that has been very extensively discussed by philosophers is that of a *substance*. But this concept has surely not provided us with any greater consensus about what is being discussed.[2] There are, nonetheless, some recurring ideas. Henceforward I shall generally talk of things, in a rather less technical sense than that usually related to substance, but in a rather more specific sense than the everyday use of the word.

In the orthodox philosophical view the things that compose the world have *properties*. A thing is typically determined as being a thing of a particular

---

[1] The relation between my earlier views and the later, processual philosophy are explained in more detail in Dupré 2024.

[2] The various senses of this term and its history from Aristotle onwards are described and discussed in Robinson 2021.

kind by its possession of a specific set of *essential* properties. This provides a solution to the pressing problem of what changes, if any, are consistent with the continued existence of a thing: it persists as long as it retains its essential properties. Suppose the essential property of a table is to include a flat surface some distance above the ground. If I paint the table it maintains this property and the table still exists. If I saw it in half, its essential property is no more, and so is the table.

Things are generally thought of as *autonomous*: they don't depend on anything else or their relations to anything else for their existence. In principle, we can imagine a universe in which the only thing that existed was my dining table. A thing is also thought of as having (at least fairly) *clear boundaries*. We know where it starts and where it ends. And finally, as I have already mentioned, a thing is stable by default. If I put my table in the attic and leave it for twenty years, I expect it still to be there. If it isn't I look for an explanation. Perhaps it has been stolen and is happily ensconced in someone else's dining room. Or perhaps it has been devoured by woodworm and is no more. I said that the account of a substance or thing was controversial, so of course not every metaphysician who sees the world as made of things requires exactly these characteristics. But if an entity does not approximate all or most of these features, its claim to be a thing becomes dubious.

Processes can initially be understood in opposition to this account of things. They are always dynamic, reflecting a broader view of the world as dynamic throughout. Hence what often requires explanation in relation to the more interesting processes is not so much the changes that they happen to undergo, though those may be important, but how they retain their stability, how it is that many of their properties stay the same. And as it turns out, the maintenance of stability is typically explained in important part by relations to the wider environment in which a process exists, so processes are seldom autonomous. And, finally, their relations to external sustaining processes are often so intimate that the process lacks clear boundaries. A process, and here I have especially in mind a living process, is typically so intimately intertwined with surrounding processes that sustain it, that there is no clear answer to the question where it ends and its environment begins.

The full articulation of a Heraclitean world is a huge project. What I will not attempt, for reasons of both space and competence, is the detailed defence of a Heraclitean physics. Fortunately, it doesn't take a lot of work to see that this has become increasingly plausible. Quantum mechanics, field theories, and much else in contemporary physics seems increasingly irreconcilable with seventeenth-century atomism. But also, showing that the world of complex things, and most especially of living things, cannot be understood in the way that atomism proposes, as merely the aggregation of the properties of constituent elements, will also show that even if atomism, somehow, were a true account of the smallest physical things, it does not support a neo-Parmenidean world view. Such a view cannot, at least, be applied to the living world, and so cannot be generally true. The explanation of the necessity of a Heraclitean view of life, and its implications, most especially for human life, will be my main task in this book.

A processual world is necessarily a world of time, so it will be worth sketching a very little history, at least as it is now told by cosmologists. The universe, we now believe, started with a bang. Before that there was nothing; but after the big bang there was still no thing; just inconceivably dense energy. Very soon—after 10^{-12} seconds, we are told—entities emerged that we call quarks and gluons. Are these things? Well, if so, they are not very stable things, as below about two trillion degrees Celsius, a temperature to which the universe had cooled after about 10^{-6} seconds, they coalesce into so-called baryons, such as neutrons and protons. Developments then proceed much more slowly. After a few minutes neutrons and protons begin to combine into small nuclei, and in a few hundred thousand years these combine with electrons to form atoms. Over many millions of years discontinuities in the distribution of matter lead to the condensation of some matter into stars, and the conditions inside stars enable the fusion of small into larger elements. And so on.

This is a story of the emergence of pockets of order out of total disorder—eddies, one might say, in the flow of mass/energy. It might be supposed that some of the entities that emerge qualify as atoms in the ancient philosophical sense. This is a difficult case to make, however. The elementary, that is, indivisible, particles, which include quarks and gluons, are not very stable at all, until they are mutually stabilized in neutrons and protons or, ultimately, atoms (in the modern sense). Most atoms are very stable, but by no means unchanging, in the short term in respect of properties such as the number and excitation levels of associated electrons, and in the long term sometimes into atoms of different kinds. Many are highly unstable chemically, tending to form more complex molecules. And they are not, of course, elementary. Descriptions of genuinely elementary particles tend to speak of intertwined oscillations rather than the stability and solidity associated with the philosophical tradition of atomism. I won't pursue this topic any further here. But, as already noted, even if some remnant of atomism can be defended at some microscopic level, its once alleged implications for the more complexly structured parts of the world cannot be sustained. It is to the explanation of this, via a Heraclitean account of the living world, that I now turn.

Concurrent with the rise of atomism in the sixteenth and seventeenth centuries was the idea that organisms, including humans, were a kind of machine. A classic example of this way of thinking can be found in the book *L'homme machine*, by the French physician and philosopher Julien Offray de La Mettrie (1709–1751), published in 1747. But the most famous exponent of the view was undoubtedly René Descartes, in whose philosophical thought it is even more famously associated with a metaphysical dualism that located the cognitive and affective features of the human within a separate, immaterial part of the person, ultimately the driver of the machine. While this alleviated many concerns about the mechanical model, notably the question of whether a machine could be conscious, it raised many equally severe difficulties. For example, the mode of interaction between the material machine and the immaterial soul or mind was extremely obscure.

In recent philosophical thinking Cartesian dualism has declined for many reasons. Among philosophers like myself, committed to a naturalistic metaphysics, the inaccessibility of an immaterial substance to empirical investigation is one sufficient reason for scepticism about such a dualism. Descartes's mechanism, on the other hand, is alive and well, and indeed has undergone a major revival in the last twenty years under the rubric of 'the new mechanism'. This movement is generally dated from an extremely influential paper in 2000 by Peter Machamer, Lindley Darden, and Carl Craver (hereafter MDC), though there are important precursors, notably Stuart Glennan (1996).[3]

MDC define mechanisms as 'entities and activities organized such that they are productive of regular changes from start or set-up to finish or termination conditions'. 'Entity' is a word often used by philosophers—including myself—to be as ontologically neutral as possible. But for MDC this is not the case. 'Entities are the things that engage in activities,' they write. They see their view as an ontological dualism: neither entities nor activities are reducible to the other. But it is an asymmetrical dualism: an activity requires an entity to perform it, whereas an entity can happily exist doing nothing. Thus, although they wish to differentiate themselves from two kinds of monists—processualists and substantivalists—they are surely much closer to the second. Their issue with the latter is whether the changes that occur in

---

[3] The other side of the dualism has now staked out its own academic territory, under the flag of consciousness studies.

the operating of a mechanism should be described simply as changes in the properties of entities, or whether, as they insist, we need to attribute some other entity—in the common, ontologically neutral sense—to the thing, its activity. This may be an interesting issue for someone who thinks the world is populated by things, but is a matter of detail to someone who, like myself, believes there is only process.

Ultimately, for mechanists, the world is an array of things. Although activities are counted among the ingredients of mechanisms, they are all inescapably tied to the entities—or let's just call them things—that implement them. Let me note, before moving on, that much of the motivation for the new mechanism is to reflect the fact that scientists very often describe themselves as looking for mechanisms, and philosophers of science naturally seek to understand what scientists think they are trying to do. But philosophers of science cannot just take the scientists' views of what they are doing—or, for that matter, simply the terms that scientists use to describe what they are doing—as sufficient grounding for ontology. We should, as naturalistic metaphysicians, try to understand what scientists have found to be the case. But how best to understand the more general outline of the world they have helped to illuminate is a distinct further philosophical task.

My main point in discussing the new mechanism is to illustrate how thing-centred metaphysics is still thriving in the philosophy of science. But I should also mention that further central features of thing-centred metaphysics are assumed by the new mechanism. First, mechanisms do the explanatory work they do from the bottom up. The entities, and perhaps the activities associated with them, are *parts* of the whole mechanism, and the whole generates the end state it does because of the behaviour and interactions of its parts. A corollary of this, to which I shall return, is that explanations of behaviour are internal to the behaving entity. Processes, as I shall explain, cannot be understood in this unidirectional, bottom-up way.

Life as Process

So now I must explain why I insist that life is entirely composed of processes. Let us recall the characteristics I listed as typically applied to things: things are by default stable, they are autonomous, they have clear boundaries, they have essential properties that determine what kind of thing they are, and, finally, they may continue to exist despite undergoing no change. I shall now show that none of these characteristics is readily applicable to typical biological entities. The obvious place to begin is with what most people take as the paradigmatic biological entity (now in the ontologically neutral sense), an organism. Think of an elephant, a mushroom, or an oak tree. While the question whether an elephant is a thing or a process will strike the philosophically uncontaminated addressee as rather odd, I suppose 'thing' is the more likely answer. Since Descartes, as already discussed, it is common to think of an organism as a mechanism or a machine, a view that remains widespread in contemporary science. A machine I take to be a kind of thing.

My most general characterization of a process above was as something that requires change to continue to exist. A table or a toaster can be stored in the attic for many years and then taken out in much the same state as it went in. The toaster may even work. Not so a duck or a mushroom. For a duck to stay alive, literally trillions of chemical reactions must take place in its body every second. Mass and energy are constantly being taken in from the environment—food, water, air—and used to drive countless metabolic processes in the cells and systems of the animal. Solid, liquid, and gaseous waste products are excreted. When this flow of energy through the system stops, we have a dead duck.[4] Physicists describe an organism as a far from equilibrium thermodynamic system. When the provision of energy stops the system returns to thermodynamic equilibrium with its environment. It is not like a car that needs externally provided energy to run, but can be left for an indefinite period doing nothing; an organism needs an input of energy to continue to exist. So, I say, the organism is a metabolic process. Metabolism refers to the full set of chemical processes that sustain its continued existence.

There is another, even more familiar sense in which an organism is a process. An organism has a life cycle. The great British biologist of the mid-twentieth century, Conrad Hal Waddington, a follower of Alfred North Whitehead, the doyen of twentieth-century process philosophy, distinguishes usefully between homeostatic and homeorhetic processes (Waddington 1957). A homeostatic process maintains certain properties of a system; metabolism is, in general, homeostatic. A homeorhetic process, on the other hand, maintains a specific trajectory. An example is the life history of an organism, but we might also point to the life histories of the cells of which the organism is composed. An organism is a developmental process. Development should not be understood, as it sometimes is, as a process

---

[4] Jacques de Vaucanson's (1709–1782) famous mechanical digesting duck could eat and excrete when made to do so by its operator, but could exist indefinitely without doing any such thing.

that leads to the steady state of the adult organism which, at some later date, begins a new process of ageing and death. There is no such steady state. Development is a continuous process, sometimes changing more rapidly than at other times, to be sure, from conception to death.

It is worth stressing the 'is' of identity in the claim that an organism is a developmental process. Often, we have a paradigm stage of the life cycle in mind when we think of an organism. If I refer to an elephant you are likely to imagine the adult behemoth rather than the cute baby elephant, let alone the elephant embryo or foetus. But all of these are parts of the same process. And though the word 'frog' is not generally used to refer to a tadpole or a frog's egg, there is no question that they can be the same organism. A tadpole becoming a frog is not a case of one being coming to an end and another coming into existence; it is one and the same being—a process—taking on a radically different form.

These two respects in which an organism must be seen as a process both contradict the attribution of stability by default. Constant metabolic activity is essential for an organism to continue to exist; without this it is dead. A developmental process, by default, is transforming into the next stage in its developmental journey; to cease to develop is to cease to be the kind of entity it is. To digress slightly for a moment, it is interesting that a few billionaires, frustrated by the fact that they cannot buy immortality, are lavishly funding scientists to attempt to stop this process and halt or stop ageing. It generally remains unclear what exact trajectory of life cycles is intended as an outcome of this project. Perhaps it is plausible, if optimistic, that the endpoint of scientific medicine will eventually lead to everyone living a healthy life for, let's say, 110 years. But the thousand-year life span that is advertised as a goal is another matter entirely. Will the developmental process be entirely halted for 900 years? Will everything be slowed down, so that people take 200 years to reach maturity? I believe that at the root of the idea, apart from sheer wishful thinking, is the machine view of the human organism. Although in actual fact there are organisms that last much longer than any machine—trees that live for several thousand years, for instance—in principle, if one can provide replacements for worn out parts, a machine could be kept going indefinitely. Unfortunately for the billionaires, we are not machines. It is not clearly impossible that humans could somehow be modified into organisms with very long life cycles. But if that were to be achieved, it would have to involve some solution to the question of the shape of that life cycle. Whether there is an attractive target here is as doubtful as the prospects of attaining it.[5]

---

[5] Fantasies of immortality will be discussed in more detail in chapter 5.

There is an even more interesting reason why all or most organisms are not autonomous. This is the near omnipresence of symbiosis (Gilbert and Epel 2015; Sapp 1994). Organisms typically depend for their proper functioning on intimate relations with organisms of other kinds. It is now well known that the human body contains trillions of non-human passengers in the form of microbes. These are especially concentrated in the digestive tract but are also found on the surface of the skin and in other body cavities. These are not, as was once assumed, merely opportunistic settlers finding a safe, warm place to live. Many of these passengers provide more or less essential services to their hosts.

To different degrees, the symbionts of multicellular organisms provide resources for digestion, the vital processing of necessary resources from the environment. Despite the omnipresence of cellulose in plants as a potential food source, the vast majority of multicellular animals are unable to digest cellulose without the help of microbes. Microbes are found in all four compartments of a cow's stomach and in its intestine, though the digestion of cellulose that they enable mainly takes place in the rumen. The roots of plants are the sites of complex communities of bacteria and fungi, as well as archaea, viruses, and protozoa, involved in the provision of resources. Fungi in these plant communities may penetrate the root and spread into the surrounding soil. The so-called mycelial webs into which these fungal processes spread may link a variety of plants, sometimes of diverse species, and are thought to benefit the entire system by transferring needed resources from one plant to another, leading some to talk of the wood wide web (see e.g. Beiler et al. 2010). In the human case, digestion, immunity, development, and possibly even cognition are affected by symbionts. As is typical of symbiotic communities, the member species in these complex associations range from essential symbionts to the parasitic and even pathological. Many species are somewhere in between, being beneficial in some circumstances, harmful in others (Méthot and Alizon 2014).

Where an organism depends on symbionts for its survival, it can hardly be said to be autonomous. It is true that some theorists have begun to argue that the true organism is the symbiotic whole, including both the multicellular organism (or 'macrobe') and all its microbial symbionts. This whole is sometimes known as a 'holobiont' (Margulis and Fester 1991). Some now take the holobiont to be the fundamental unit of evolution (Zilber-Rosenberg and Rosenberg 2008). While this no doubt helps address the problem of autonomy, it immediately raises a different worry.

A further feature of traditional conceptions of the thing is that it should have reasonably clear boundaries. But this is greatly problematized by the kind of symbiosis I have been discussing. I noted that symbiotic microbes range from essential mutualists or even, perhaps, organisms from which the holobiont gains benefits and returns none, and so might be seen to parasitize, to parasites and pathogens damaging to the whole. Where in this range do we draw the line between organism (holobiont) and other? It is hard to see a well-motivated answer to this question. French philosopher Thomas Pradeu has suggested that the immune system can provide an answer to this question (Pradeu 2011). The organism, he proposes, is composed of just those parts that are tolerated by the immune system. There is much to be said for this proposal, but I won't try to assess its difficulties and advantages here. I will just note that the organism distinguished by this proposal is a surprising one. Not only will it contain some, but not all, of an organism's symbiotic microbes, but there will be parts of the macrobe that are not included, as much of the immune system's work is the disposal of malfunctioning or dying parts of this.

The case briefly mentioned of plants connected into one system by a mycelial web provides an extreme version of the problem of boundaries, since it may appear that a number of taxonomically diverse plants are all part of the same holobiont. Plants have anyhow for a long time been recognized as providing a difficulty for clear discrimination of individuals because of the difficulty of distinguishing growth from reproduction. Many plants produce, for instance, what look like new individuals that grow from spreading roots. The new plant, since resulting simply from the growth of the old, does not appear to be a genuinely new individual, but can easily be made to do so by severing the root that connects the parental individual to its vegetative offspring. Gardeners may perform such an act of separation and transplant the severed part to a distant location. Is this a form of reproduction by root severance, or is it just a case of individuals with relatively arbitrary boundaries? The difficulty of finding boundaries around biological systems will recur at several points in this book.

Essences

I must now say something on a more technical philosophical point, the question of essence. It has very widely been assumed that a thing must be distinguished by an essential property or set of properties, the essence of the kind of which it is a member. Things may, of course, belong to many different kinds, so there is a question of which kind it essentially belongs to. An answer to this question seems to require a privileged set of kinds to exactly one of which everything belongs. I think this is in fact an impossible need to satisfy, but I won't pursue exactly this problem now (but see Dupré 1993). Other considerations will more usefully show the difficulties with this postulation of essences.

The essence of a thing serves various philosophical purposes, of which two are the most important. First, as just noted, it determines what kind of thing it is; it is, in John Locke's memorable phrase, 'the very being of any-thing, whereby it is what it is' (Locke 1689/1975, III. iii. 15); second it tells us when a thing comes into or goes out of existence: it exists if and only if it possesses its essential property. Philosophers addressing these problems have argued that their solution requires that there be an essence of a thing. Typically the examples used are of biological organisms.

Unfortunately, over recent years philosophers of biology have insisted that there can be no essences of biological kinds (see e.g. Hull 1967; Dupré 1993). The simple reason for this is the fact of evolution. Evolution by natural selec-tion requires variability within a kind that evolves, as it requires variation in fitness. Evolution could occur if variation were restricted to non-essential, so-called *accidental*, properties. But evolution is also expected to explain the transition from one kind to another. Indeed, if biologists are correct that all currently existing organisms share some common ancestral population, it follows that there is a series of evolutionary transitions linking any two kinds. Suppose we want to trace this series from a frog to an oak tree, say, we simply go back through the frog's ancestral lineage until we come to the common ancestor with an oak tree, and then forward from there to the oak tree. All along this pathway kinds with one essence would have had to evolve into kinds with a different essence, and in doing so there is seldom a clear point in the transition from creatures with the purported essence of the former kind to creatures with the essence of the latter. And this, finally, contradicts the ini-tial assumption that everything must have the essential property of its kind.

If, indeed, there is no essence to a thing, how are we to address the ques-tions that essences were invented to answer. First, if there is no essence to an organism, how are we to know what kind it is or, more specifically, what species it belongs to? The answer should be obvious from the preceding ar-gument from evolution: there need be no clear boundaries between species and sometimes we will not be able to assign an organism unambiguously to a particular species. I cannot go into any detail here about the debates over the process of speciation,[6] but however this may turn out, it seems inescapable that during this process there may be borderline cases which, at the very best, can only be unequivocally assigned to the new or old species retrospectively.

In fact, we should welcome this conclusion for other reasons. One is the frequency of hybridity. Although it is still widely asserted, even by biologists, that a species should be defined as a closed group of interbreeding organisms, hybridization between related species is very common (Mallet 2005); but a hybridizing species is not a reproductively closed group, and a hybrid can be assigned to either of its parental species. A phenomenon even more problematic than hybridization is lateral gene transfer, which is very common in microbial life, but can happen in all parts of the living world. This is the movement of genes between sometimes distantly related organisms. There are several processes by which bacteria achieve this result; among multicellular organisms it can be mediated by viruses. In the latter case it is unlikely to threaten the species membership of an organism, but in the case of bacteria it creates serious problems for classification. Classification of bacteria is generally based on particular parts of the genome that are thought to be highly conserved in evolution. But if genes can be acquired through lateral gene transfer, the genes in a bacterium may come from many different sources, and the assignment to a species may depend on the choice of gene for classificatory purposes. This is incompatible with the idea that species membership is defined by an essential property.

This last point is an example of a more general problem that is by no means limited to microbial life. A growing body of philosophical and biological work has come to the conclusion that there is no unique way of defining the species, assumed to be the ground level unit of a taxonomic scheme for biology. Moreover, it seems that different definitions of the species may be suited to different biological purposes, for example evolutionary, ecological, or functional. This conclusion is incompatible with the existence of a privileged taxonomic scheme defined by essences.

The second purpose just mentioned for assigning or discovering essential properties is for determining when entities come into and out of existence. For paradigmatic things it seems a reasonable demand that we should be able to answer this question. It would be surprising if one were unable to say unequivocally whether a table continued to exist. As noted earlier, painting

---

[6] For state of the art discussions of these debates, see Wilkins, Zachos, and Pavlinov 2022.

the surface or replacing a wobbly leg seem to pose no threat to its survival, whereas sawing it in half surely does. A plausible criterion is that a table survives as long as it remains able to serve its basic function, keeping objects of various kinds stably positioned above the ground.

A famous problem with this simple view is posed by the ship of Theseus. Theseus, as the story goes, replaced damaged or decaying parts of his ship until eventually every part of the original ship had been replaced. However, all the replaced parts were carefully stored, and when the last part was replaced the original parts were reassembled into a second ship. Which is the original ship of Theseus? The one that has a continuous history of functioning as a ship, or the one that contains all the surviving material substance of the original ship? A similar history is easy to describe for our table, and suggests that there are two possible criteria: a material one and a functional one. It is worth remembering here that most of the material composition of an organism is replaced over periods from hours to a few years.

My own favoured response to this problem is indifference. Why should there not be two legitimate ways of tracking the history of an entity? If someone objects that this is a contradiction: if ship A is identical to both ship B and ship C, then ship B is identical to ship C, which is impossible, I reply merely that the material ship and the functional ship are not the same *process*. One thing may be different processes, as the organism is both a developmental and a metabolic process. The problem may seem more pressing in the philosophically more contentious case of a person. In what sense am I the same person as existed many years ago as a child? This problem I shall defer to chapter 4.

I began this chapter by sketching some current scientific thinking on the early history of the universe. This was a story of the gradual emergence of increasingly complex structures from a pure and undirected flow of energy. This is, of course, exactly how we think of the evolution of life over the past few billion years on our own planet. The details of that process must wait for the next chapter, but for now let me review some of what I have been saying about how we should understand this universe in which, as Heraclitus is supposed to have said, 'everything flows'.

One general answer I have considered is in terms of a dualism of processes and things. The first quark, 10^{-12} seconds after the big bang, was the first thing. Later followed hadrons (neutrons and protons), atoms, molecules, stars, planets, and, eventually, organisms. But we have seen that these have diverse degrees of stability, autonomy, and, as will become clearer when we turn to more biological detail, clarity of boundaries. Isolated quarks, whatever their boundaries, last for no more than about 10^{-25} seconds at temperatures below two trillion degrees Celsius, and are otherwise inextricably conjoined with other quarks and gluons forming hadrons, such as neutrons or protons. Their stability is at the cost of any autonomy. Stars, at a rather different order of magnitude, last a long time, but have characteristic life cycles as their internal energy sources are gradually extinguished, meanwhile changing their material composition, and eventually ending as supernovae, black holes, red giants, or white dwarves, depending on their size. Perhaps there are some reasonable candidates for things among this motley crew, but there is a far more satisfying, because more general, picture available. This is the view of the world for which the eddy in a flowing stream provides the paradigm.

Eddies can be very stable. Think of the red spot on Jupiter. This has been observable on the giant planet since sufficiently powerful telescopes have existed to observe it, perhaps 500 years. This is in fact an enormous storm—its width is about three times the diameter of the Earth—essentially an eddy in the flow of atmospheric gases. Its structure is maintained by the rapid winds that circulate around it.

This brings me back to perhaps the most significant practical implication of a processual world view. In the standard view, the persistence of a thing is the default assumption. Scientific investigation is called for to explain when things change, or when they come into or go out of existence but not, typically, to explain their continued existence when nothing happens to them. But the persistence of a process always requires some explanation. Indeed, I defined a process as something that requires activity for its continued existence. Persistent structures in a world of process require stabilization, and to understand these structures we must know how they are stabilized. Brief reflections on the importance of metabolism and symbiosis quickly show how this applies to biological organisms. It seems to me that detailed explanation of chemical bonding and physical forces also includes a similar kind of dynamism, but my own focus is on life and if students of the physical sciences insist that this is not the case for their particular areas of interest, so be it. From now on I shall be concerned only with life and life, I claim, is always and everywhere processual.

Some Philosophical Background

I shall conclude this chapter with some brief remarks on some well-trodden philosophical background. Some philosophers like to divide entities into continuants and occurrents. Peter Simons explains:

> There is a pervasive and basic duality in the way in which we speak about
> objects in the real (spatio-temporal–causal) world. Some things are in time
> in such a way that they grow through the accumulation of parts—temporal
> parts—as time goes by. These are events, processes, and states. The other
> sort of things are those traditionally called substances: people and other
> animals and organisms, houses, cars and other artefacts, mountains, rivers,
> planets and stars, and other natural objects. Unlike the other basic sort,
> they do not add temporal parts but are said to remain identical or the same
> at different times. It is the same person, cat, car, or river it was yesterday,
> notwithstanding changes to it in the meantime. I . . . call the second group
> continuants and the first group occurrents. (Simons 2018)

So processes are occurrents and things or substances are continuants. I like
to use rivers as an example of processes, but Simons uses them as an example
of continuants, so clearly I must reject this alignment. In fact, I propose that
what we generally speak of as things are continuant processes. I have certainly heard philosophers complain that this was a straightforward contradiction: continuants, they say, are substances not processes. Fortunately
some philosophers do now defend the existence of continuant processes,
though not perhaps in quite the sense I intend.[7] I do insist, anyhow, that
many, perhaps all, familiar continuants exhibit crucial features generally associated with process rather than with substance. I claim that some processes
are occurrents, some are continuants. But everything changes.

Note that the quote from Simons begins with a claim about how we speak.
The second sentence, however, appears to be a statement about the world,
about the nature of the entities we think of as events or processes. My argument is that the second category, of substances, in fact shares the feature that
Simons attributes to processes. They may indeed 'be *said* to remain identical
. . . at different times' (my italics), but this is a matter of how we speak, not
of some ontological difference between these and the entities we speak of as
processes. I'm not sure that Simons would disagree very strongly with this, as
in the paper cited he agrees that process is ontologically more fundamental
than substance.

The final part of Simons's remarks, concerning the sameness of a continuant over time, is sometimes expressed in the idea that a continuant is

---

[7] E.g. Stout 2016. Stout would not, I think, support the extension of the category of continuant processes to the extent that I propose, but does defend the general intelligibility and
application of the category.

'wholly present' whenever it exists, whereas occurrents have temporal parts, only one of which is present at any moment. I am suspicious of the idea of being 'wholly present'. What is it supposed to contrast with? If I am watching a football match, I am wholly present and so, I think, is the match. Of course, the winning goal may not yet have happened; likewise my heart attack that it provoked, following which I was fully absent. It is true enough that events such as football matches and concerts are frequently described in terms of their temporal parts. But so might we describe a frog. Is it wholly present, given that the tadpole and the egg from which it came no longer exist? I shall return to this topic in chapter 4.

A number of philosophers have objected to the position I take in favour of a broadly neo-Aristotelian account of substance (Austin 2020; Morgan 2022; Steward 2020; Skrzypek 2023), a version of hylomorphism. Hylomorphism considers individuals (individual substances) to be composed of both matter and form, and form here includes, at least for the case of organisms, a dynamic element corresponding to the organism's life cycle. My general response to this is that what distinguishes such views from processual views, if anything, is the essential character of the appeal to form. I reject any attempt to define essential properties for organisms, for reasons that are well known in philosophy of biology (Dupré 1993), and which I shall elaborate at various points in the book. But, as the preceding sentences suggest, I do not see the divide between processual and neo-Aristotelian positions to be very wide. The former do, certainly, recognize the dynamic nature of life. They do also, however, hold on to notions of substance and essence which, I think, we are much better off without.[8]

Philosophically sophisticated readers will be aware that the preceding few paragraphs hardly touch the surface of the more technical debate over the competing merits of substance and process ontologies. However, it would be well beyond the scope of this book to provide an in-depth account of these debates.[9] My goal here is rather to sketch the arguments for, and advantages of, a process biology.[10] In what follows, I hope to show that a wide variety of traditional problems in biology and philosophy look quite different from a processual perspective, and that this perspective offers attractive ways

---

[8] A detailed response to neo-Aristotelian accounts of living systems, as well as an excellent account of the metaphysical history and rationale for process biology, is provided by Meincke 2023.

[9] I go into more detail on these topics in Dupré 2021a.

[10] Various perspectives from both sides of the debate can be found in Meincke and Dupré 2021.

forward with addressing and even sometimes resolving them. The book, then, rather than addressing directly arguments for and against this process perspective, might be seen as an extended abductive argument for a process philosophy.

In this chapter I have argued that we should reject the ontology of stable, hard-boundaried things that has dominated our thinking since at least the scientific revolution, and embrace instead an ontology of process and fluidity, one in which what we traditionally think of as things are merely temporary pockets of more or less orderly structure in the surrounding more or less disorderly flow. In the remainder of this book I will spell out some further implications and applications of this view, first in relation to evolution, and then in relation to various aspects of the human condition. I hope that in combination with the arguments so far presented for the philosophical correctness of the processualist position, the illumination it can provide on these vital problems will persuade even the more sceptical reader to merge into the growing processualist stream.

2

Evolution and Lineages

Introduction

In the previous chapter I described a contrast between a philosophical ortho-doxy that sees the world as composed of things and an alternative view that sees everything as process. What we tend to think of as things are, I claim, no more than temporarily stabilized structures in a world of change or process; I illustrated the point with a discussion of the nature of the organism. In this chapter I turn to the most celebrated scientific idea in biology, evolution. No one doubts, of course, that evolution is a process. Evolution is a theory of change, of how the vast diversity of our current biosphere came into being from vastly simpler and more homogeneous origins. Nonetheless, a proces-sual perspective can have profound implications for our understanding of evolution.

Let me begin by getting clear about what it is that evolves. We sometimes say, casually, that humans evolved. But of course, no individual human evolved. Humans develop, mature, and eventually die. But they do not evolve. What evolved is the lineage which humans now constitute. But there are some subtleties to the concept of a lineage that I want now to explore.

It will be helpful to start with the very familiar image of the tree of life (figure 2.1). Let us assume for the time being that all species are sexual, and that there is no hybridization. The tree of life represents the history of life. It is widely agreed that all living things (with the possible exception of vir-uses, which many biologists consider not to be living anyhow) derived from a common ancestor, sometimes known as LUCA, the last universal common ancestor. The tree of life can be drawn at many different scales. In figures 2.2 and 2.3 we see images with increasingly fine detail of human evolution, ei-ther of which would be a very small detail of the element labelled 'animals' on a larger-scale tree of life.

Figure 2.2 shows the evolutionary history of primates, figure 2.3 of hominins, the one extant and several extinct human species, with chimpan-zees for reference. A complete tree of life—certainly not something we will

Everyone Flows. John Dupre, Oxford University Press. © John Dupré 2025.
DOI: 10.1093/9780198941866.003.0002

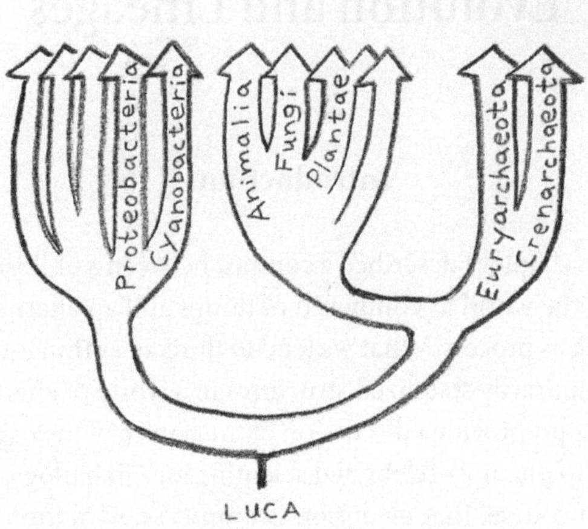

Figure 2.1 This is a highly simplified picture of the evolutionary tree of life, showing the three domains of life, Bacteria, Eukarya, and Archaea, with some major representatives of each domain. The figure is a simplified version of figure 2.4, which shows the complexities added by various forms of gene flow. (G. Anderson 2024.)

ever be able to draw—would have all the detail of this third picture (which still excludes several known and many as yet unknown extinct species) on every part of the map. The fuzzy lines on figures 2.2 and 2.3 are reminders of this largely unknown diversity. The lower parts of the first, large-scale map would be covered with great sub-trees of now extinct species. But note that however much we simplify the tree by increasing its scale, any line on it represents what was once a species. If we look at lines near the bottom of the large-scale picture, the length of line that remains the same species is very short. All the branches that are not shown represent points where the species might end. For cladists, now the dominant school in taxonomy, a branch is by definition a point at which one species ends and two others begin. In reality, nonetheless, there are continuous lineages from LUCA to every later

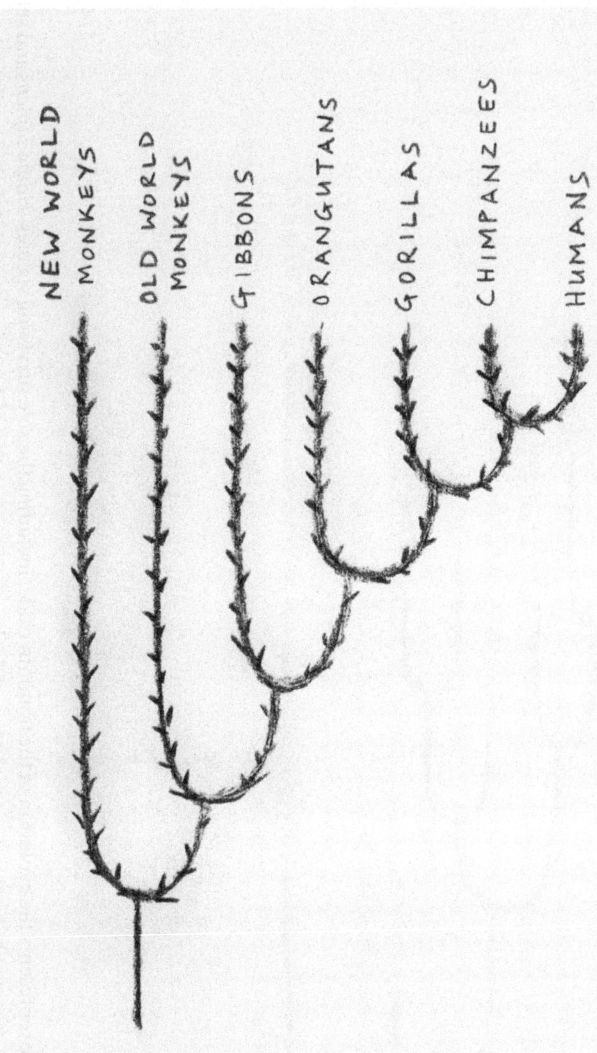

Figure 2.2 This image shows the main divisions in the primate clade. The branched lines indicate the many, often unknown, extinct side branches. (G. Anderson 2024.)

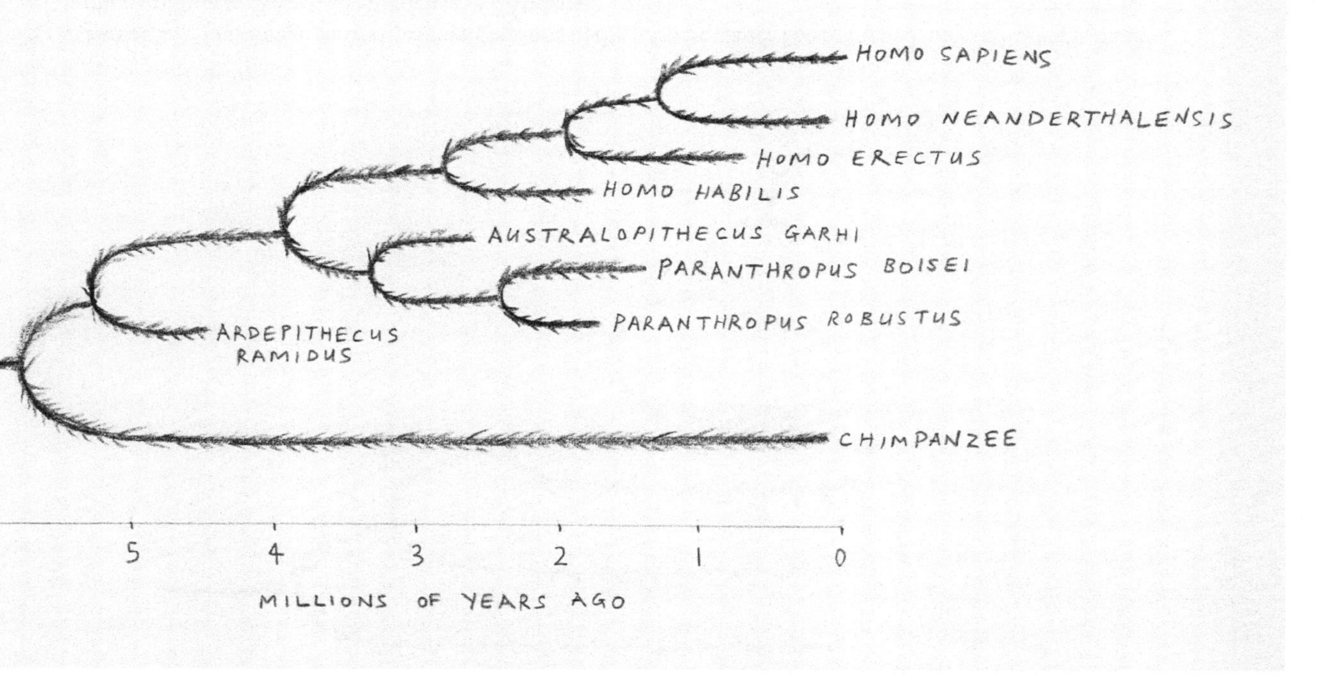

Figure 2.3 This image represents the evolution of the hominin clade, including the sole surviving species, *Homo sapiens* and selected extinct species. The diagram includes the closest relatives of the hominins, genus Pan (Chimpanzees and Bonobos). (G. Anderson 2024.)

species; how these lineages are divided into distinct species is largely a conventional matter.

With the tree of life in focus we can now introduce two key concepts that are sometimes confused, the lineage and the clade. The clade, which has become an entity of central interest to taxonomists, classifiers of living diversity, is a species and all its descendants. At whatever scale we draw the tree of life, if we cut through any of its branches what we thereby detach is a clade. The clade, then, is forward looking, from a species to all its descendants. Cladists believe that every taxonomic unit should be a complete clade—hence the necessity that a species be said to cease to exist when it splits into two.

Lineages, however, are backward looking. A lineage can be found by starting at a species and tracing back through the sequence of species that preceded it, in principle as far as LUCA. I noted that lineages and clades are sometimes confused. If you search on the internet for images of a biological lineage, you will find mainly pictures of clades. Definitions of the lineage are often forward looking, though sometimes including a word such as 'line' that points to a more accurate understanding. But the backward-looking definition is far more straightforward. Going backwards from the outer reaches of the tree there is never any question which way to turn—you should always go backwards in time.

A significant consequence of thinking of a lineage this way is that at every point the lineage is precisely coextensive with a species. This is also why it is useful to restrict the account in the first place to sexual species. A sexual species, by virtue of actual or possible reproductive links between its members, can be seen as a coherent and persistent process. Hence the lineage as an ancestral/descendant sequence of species can be seen as a continuous process. Evolutionary change, finally, can be seen in the differences between the properties of the species at one stage of the lineage and the properties of the species at a later stage. There need not, of course, be any sharp transitions between species. For a strict cladist, a new species appears simply because the lineage encounters a new branch in the tree of life. It may be that the origination of a branch requires some kind of significant change; but that may be located in the other branch.

At this point one might relax the assumption of no hybridity and even, relatedly, allow lateral gene flow by other means, such as viral vectors. Lineages are often thought of as vehicles for gene flow, but introgression of genes from other sources is not necessarily a problem; indeed, it may be seen as a significant potential origin of evolutionary change. Hybridity presents a slightly different problem. Suppose that two closely related species have a lot of gene flow between them through production of fertile hybrids. Why should we think there are two species rather than one? The answer is that there is no reason. Fortunately, this is not a problem for a process ontology. Processes do not require the sharp boundaries assumed for things, and counting of processes may often be a matter of arbitrary stipulation. In the present case, history may eventually decide the matter as the lineages do or do not diverge. Retrospectively, there may be no question *whether* they have become distinct. But there may be no objective answer to the question *when* they became distinct.

The promiscuous lateral gene transfer found among many bacteria points to a more serious problem. Figure 2.4 shows an image of the tree—or better net—of life, derived from a well-known image from the evolutionary microbiologist Ford Doolittle.

As well as lateral transfers between cells, represented by horizontal black lines, figure 2.4 includes the evolutionary histories of mobile genetic units, most importantly viruses and plasmids, dependent on cells for their reproduction, but often capable of migrating between cells of different kinds and

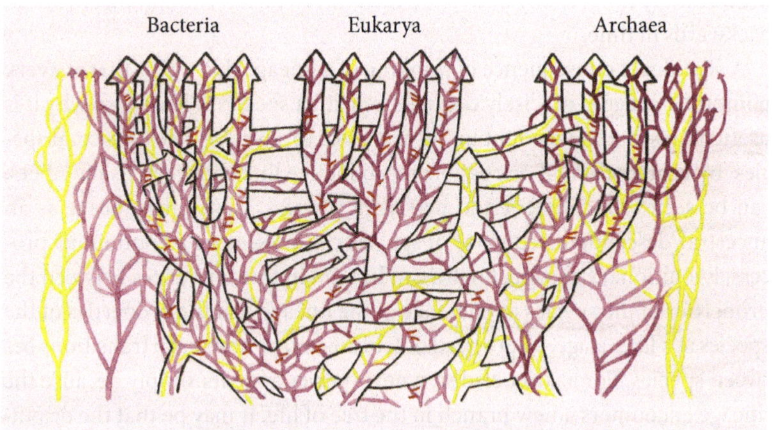

Figure 2.4 In this elaboration of figure 2.1, the black lines in this image represent cellular lineages. The net-like rather than strictly tree-like form indicates major events of lateral gene transfer. Purple and yellow lines represent, respectively, the evolution of viral and plasmid lineages, including both vertical and lateral processes. Red lines represent transfers between cellular organisms and mobile genetic elements. This is adapted from a well-known image from Doolittle 1999. The present version is from Bapteste and Anderson 2018. (G. Anderson, watercolour on paper, 2015.)

of adding their genetic material to that of the cell lineage. These additions provide a much more realistic view of the complexity of the genetic relations presented in simplified form in figure 2.1. It also, incidentally, explains the attraction of the idea that much novelty in evolution might arise from evolutionary mergers rather than, as generally assumed, genetic change internal to a cell lineage (Sapp 1994; Margulis and Fester 1991). The most famous application of this idea is to the widely accepted account of the origin of the eukaryotic cell from the merger of prokaryotic (bacterial and archaeal) precursors (Margulis 1981).

Note that in light of the structure represented in figure 2.4 there is no unambiguous way of tracing the lineage back from a point on this tree as there are mergers as well as splits in the course of evolution. Putting the same point in a different way, lateral gene transfer means that even knowing the entire direct cell-line ancestry of a cell will not provide complete information about its genetic ancestry. Moreover, since micro-organisms are asexual, it is anyhow much less clear what constitutes a lineage in the parts of the tree composed of these. Any cell seems to have an equal claim a priori to be the founder of a new clade; and many cells will be parts of large numbers of lineages. As a result of these problems there is considerable controversy whether there are coherent lineages among bacteria and other asexual, single-celled organisms. Once again this is not a problem for a process ontology. We have a mass of cells dividing and mutating and transferring genetic material. There is strongly stabilized structure at the level of the cell and the single-cell lineage. At the level of the species lineage there may be none. Some scientists have recently suggested that most bacteria do belong to species-like groups exchanging genes very largely among themselves, so-called homologous recombination (Bobay and Ochman 2017). If so, there are indeed real species-like lineages. This is an interesting scientific issue but does not really touch my present concerns.

The Ontology of Evolution

I have already addressed the first question that arises in giving a proper account of evolution, What evolves? The answer is lineages, species, or sequences of species. Evolution occurs when distribution of traits in a lineage at one time differs from the distribution at a later time. Are species evolving lineages? With one minor qualification, yes. Many species consist of a number of more or less isolated populations. The rabbits in Australia, which may still count as belonging to the same species as their European ancestors, may not be part of the same reproductively linked community, though they no doubt still maintain at least possible reproductive links. We may treat this problem much as we treated hybridizing species. If the species stay apart indefinitely, they will surely diverge to the point that they become different species. If they reunite, then they either will or will not remerge. So retrospectively we can decide the question. At the present time, we may say what we like.

The reader will have noticed that I have been speaking of species as a kind of entity, in fact, in my view anyhow, a continuant process. But it is also common to think of a species as a classificatory or kind term. 'Rabbit' one might think was a general term of which Peter Rabbit, Roger Rabbit, and Flopsy Bunny are singular instances. So can it also name a singular entity, the process that is the rabbit lineage? In fact, as I shall explain a bit later, I think it can. But a philosophically less puzzling proposal is to suggest that what Peter, Roger, and Flopsy have in common is not being members of a kind, but rather being parts of a lineage. This also accords with the fact that given the points noted in the previous chapter about evolution and variability, it may be difficult to say what the criterion could be for membership of the rabbit kind. As I argued in the last chapter (and in much more detail in earlier work (Dupré 1993)), at least there is no essential property of rabbits that could serve this function. For now, anyhow, let me say more about lineages, including species, as individuals.

Species as Individuals

Philosophers of biology will know that the claim that species are individuals was a notorious thesis, proposed by the biologist Michael Ghiselin and the philosopher David Hull in the 1970s and eventually widely accepted (Ghiselin 1974; Hull 1976). One reason I say 'notorious' is that philosophers, at least, had taken terms like 'rabbit', 'horse', or 'oak tree' as paradigmatic kind terms for over two millennia; to be told that they were proper names of individuals seemed rather shocking. On the other hand, some evolutionary biologists were quick to point out that this was what they had always believed. Certainly, earlier accounts can be found that accord well with this view. In 1961, for instance, the renowned evolutionist George Gaylord Simpson had written that a species is 'a lineage (an ancestral descendent sequence of populations) evolving separately from others and with its own unitary evolutionary role' (Simpson 1961), very much the view I have just described.

But there is a second reason why the species as individuals thesis seemed incredible to some, and that was that a species seemed such a peculiar kind of individual. The parts of a species are entirely disconnected from one another and move independently from one another. Apart from the partial and patchy network of reproductive links they seem in no way integrated with one another. And of course, all these problems are exacerbated if we allow a species to include a number of isolated populations, like the rabbits in Europe, Australia, and elsewhere.

This second problem is quickly solved, however, when we recognize that a species or a lineage is not an individual thing but an individual process. What we need here is the concept of a continuant process that I outlined at the end of the last chapter, and which is sometimes ruled out by traditional philosophical definitions. There are many examples of processes that consist of spatially discrete, more or less autonomous parts. Think of a battle, or a coastal erosion. The separation of a river into distinct streams parallels the separation of rabbits between continents. Perhaps they will rejoin and continue to be parts of the same river; perhaps they will take their own separate ways to the sea and be judged to be separate rivers. Individual processes need not have the integrity or sharpness of boundary generally required of individual things, but they may have, nonetheless, the degree of integrity and distinctness appropriate to a process.

Having said what I think a species is I must enter two important qualifications. First, I don't think that *all* species are processes. As I have already noted, it is highly questionable whether bacteria or other microbial organisms form proper lineages. If they do not, we still need to find ways of classifying them, and cladistic methods will not be available. This brings me to my second, more general point. The word 'species' has, for centuries, been used to name our so-called basal units of classification, or basal taxa. This is the lowest level of classification that is considered biologically significant (divisions into subspecies, races, and suchlike are generally considered of minor biological importance) and hence the answer to the question, What kind of organism is this? Surely, we should not abandon this usage because the word 'species' has come to have a more specific theoretical meaning as well. It is convenient, as is no doubt often the case, if the organisms that form the parts of a lineage coincide exactly with the organisms that fall into the same basal taxonomic unit. But we know from the case of bacteria that this cannot always be the case, as sometimes there is no well-defined lineage to map onto the taxonomic unit. And in fact, given the very different roles that these uses of the word 'species' play, we should not be surprised if there are other occasions where their referents diverge.

Ernst Mayr, another leading figure in twentieth-century evolutionary biology, provided probably the most widely known definition of the species, the so-called biological species concept (BSC) (Mayr 1942). According to Mayr a species is a group of organisms connected to one another and isolated from other organisms by sexual relations and thus gene flow. Mayr did not much care for microbes and since he didn't think they formed species they didn't provide any problem for his preferred concept of a species. But this is, of course, just to assume the priority of evolutionary theory over the pragmatic demands of taxonomy. In fact, it is best to read Mayr's criterion as identifying species individuals. Sexual reproduction does indeed seem often to be the key factor that provides coherence to such individual processes when they exist. But, as I have insisted, there is no reason to assume that every organism is part of a species in this sense, as opposed to being a member of a species in the taxonomic sense.[1]

One may usefully think of the relation between the cladistic conception of the species that I have mainly assumed and the BSC as that between a process extended in time and a three-dimensional snapshot of that process. The BSC says what it is to be a member of a species at a time, whereas the cladistic account describes the extended process that is constituted by all such members past, present, and future. As I have noted, it is sexual reproduction that most decisively provides the boundaries of a species, so the fit between these two conceptions, or perspectives, is no surprise.

I will make just one further point on this topic. Species, I have said, can be either units of evolution or units of classification. Often the two coincide, *mutatis mutandis*. However, classification is a highly pragmatic activity. It is essential for the transmission and storage of all manner of biological information. If I see an animal behaving in an interesting way, I cannot communicate this information usefully without knowing what kind of animal I am observing. To serve this purpose there is an optimal size of the basal taxon. 'Mammal' would not serve this purpose well, being far too broad; but nor would the division of the capercaillie into seventeen distinguishable

---

[1] As so often in biology, we should not assume sharp boundaries, in this case between sexual and asexual reproduction. In protists (single-celled eukaryotes), sex seems to be rare, but many species have the capacity for sexual reproduction. As noted in the text, even prokaryotes exchange genetic material in ways that serve to preserve lineages.

subspecies. Evolution, on the other hand, has no such prejudices. Lineages may diverge into a large number of fully distinct sublineages, or they may last for very long periods of time, changing dramatically or becoming morphologically highly diverse. Classification and evolutionary theory aim to track different phenomena; on the one hand clusters of behavioural or morphological similarity, on the other the divergence or continuity of evolving populations. Insisting on their coincidence may often lead to a very inferior classification. Every organism should, ideally, belong to a classificatory species; many will not be part of a coherent lineage. Perhaps it would be good to have different words for these very different concepts. But often the members of one and the parts of the other coincide. And, surprisingly or not, the polysemic character of the word 'species' doesn't seem to do much harm.

Stabilizing Species

I want now to say a bit more about these species processes and how they evolve. In the previous chapter I noted that a fundamental distinction between processes and things is that whereas the latter are assumed to be stable by default, for the former, whatever stability they show requires explanation. So what explains the stability of a biological lineage? A lineage, in the first place, involves the constant production of organisms fairly similar to those extant at the time of production. This is clearly a necessary condition for the persistence of the lineage through time. But to count as an individual process, rather than merely a set of disconnected events of organism production, a species must exhibit some kind of internal organization. For sexual organisms this may be sufficiently achieved by reproductive links; reproductive connection and isolation from the introgression of alien genes are often taken as the defining characteristics of a species. For many species—notably our own—there are also social interactions that provide further internal organization. A set of reproductive and perhaps social links between members may seem a rather paltry degree of organization compared to the exquisite organization of an animal. But as David Hull writes:

> Most organisms do exhibit more internal organization than most species, but this difference is one of degree, not kind. Most species do not exhibit the internal organization common in vertebrate organisms, but the same can be said for plants as organisms. Most plants do not exhibit the internal organization common in vertebrate organisms. (Hull 1999, 32)

So supposing we allow that a lineage displays sufficient organization to qualify as an individual process, what sustains and stabilizes this organization? The first and most important answer to this question is natural selection. A process perspective highlights the importance of the often-underappreciated point that most natural selection is stabilizing. Why do we have a more or less coherent lineage, constantly generating organisms of a more or less similar kind? Because the great majority of substantial changes to the organism reduce its fitness. Because the organism is an exquisitely organized process, closely aligned with the environment in which it lives, most changes disrupt this organization in some way. Given that most species produce offspring greatly in excess of the numbers required to maintain themselves, there is plenty of opportunity for competition, and the novel will generally lose out. So natural selection alone may be sufficient to stabilize a coherent species process.

The importance of stabilizing selection tends to be underappreciated, partly because the idea of natural selection as a creative force often seems more exciting, and perhaps also partly because the philosophical preference for things over processes tends to obscure the great importance of stabilizing processes. Just to emphasize the point, imagine a world in which survival was always purely a matter of chance. Individual lineages would gradually drift away from their starting condition, so that even if, somehow, we started with a reasonably homogeneous and coherent species, the homogeneity would decrease generation by generation. To use a phrase dating from Georges Cuvier (1769–1832), but also used frequently by Darwin, natural selection ensures that the lineage remains within the limits of its 'conditions of existence'.[2]

I should say a bit more about the factor that Ernst Mayr took to be the defining factor for a species, *sexual* reproduction. To connect with the last point, stabilizing selection is plausibly a necessary condition for sexual reproduction. Without it, variation within the species would inevitably increase, sexual compatibility would decline, and sexual species would eventually become extinct. Perhaps a similar situation obtains in much of the microbial world, where in the absence of sexually contained lineages, there is competition between evanescent clades, and no persistent individual lineages emerge (though note again the possibility of quasi-sexual bacterial lineages (Bobay and Ochman 2017)).

---

[2] See also Reiss 2009.

One way of seeing the importance of sexual reproduction is that it maintains the boundaries between processes. As we saw in chapter 1, processes do not necessarily have sharp boundaries and are often liable to merge or intertwine. But without any limitations to gradual merger with surrounding lineages, it would be impossible to maintain a complex and coordinated suite of adapted traits. Sexual reproduction can function in the way Mayr envisaged to construct and enforce the boundary between lineage processes.

Given that the boundary between related processes is by no means a given, it is no surprise that evolution should have provided lineages with various means of enforcing their boundaries. Adequately isolated species survive longer than those that dissipate into their living surroundings. Natural selection, as Richard Dawkins insightfully noted, is at its most general the survival of the stable,[3] and to be stable a process must have some kind of boundary. The boundaries devised by sexual species are an amazing biological phenomenon. A great variety of isolating mechanisms serve to decrease the likelihood of mating with an organism external to the lineage. Isolation may be achieved ecologically, behaviourally, or mechanically, as in the lock and key morphology of many insect genitalia. Isolating mechanisms may be temporal, via desynchronized life cycles, for example, adult insects or tree pollen appearing at a time distinct from that favoured by similar species. If these devices should fail, various forms of genetic incompatibility and hybrid sterility that come into effect after mating provide a second layer of border defence.

There is also a positive analogue of these isolating systems, so-called mate recognition systems. The exotic colour patterns of many reef fish, the elaborate songs of birds, the signalling molecules of many fish and insects, the flashing lights of fireflies, the extraordinary dances of birds of paradise, all likely serve mate recognition purposes. The function of such traits is often difficult to pin down, as such signals are very likely to provide a basis for intraspecific competition for mates and may even trigger processes of runaway sexual selection. But the importance of mate recognition is not in question. Indeed it has even been considered so important as to provide the best way to define the species (Paterson 1985; though this seems an exaggeration).

Boundaries between species, nonetheless, are seldom sharp. Hybridity is common in most domains of life. Depending on the fertility of hybrids, genes

---

[3] As Dawkins writes in his famous book, *The Selfish Gene* (1976), 'Darwin's "survival of the fittest" is really a special case of a more general law of *survival of the stable*. The universe is populated by stable things.' I am less enthusiastic about some other aspects of this highly influential work.

may flow freely between species. A classic study on American oaks showed that despite extensive gene flow, species retained strong morphological distinctness (Van Valen 1976), suggesting that natural selection, tracking distinct fitness maxima with a fitness trough separating them, may often be sufficient to counteract the homogenizing effects of gene flow. In other cases it may be a matter of judgement whether there are distinct species at all. That is often how it is with processes.

A final crucial way in which lineages are stabilized is niche construction. Organisms do not merely track the environment, maintaining sufficient adaptation to the current state of affairs. Their activities change the environment and actively construct the environments to which they are adapted—indeed, very little in biology is one-directional and linear in the way that passive tracking of the environment proposes. Dramatic examples of niche construction are such structures as the beaver dam or the termite mound, but all or almost all organisms make some changes to their environment, and hence to the environment to which they are better or worse adapted (Odling-Smee, Laland, and Feldman 2003). A nice case discussed in detail by no less a biologist than Charles Darwin is the earthworm (Darwin 1881). The earthworm evolved from an aquatic ancestor and never fully adapted to life in a dry environment. However, by cycling organic material into the soil earthworms are very effective at maintaining the moisture level in the soil to a level at which they can flourish. As all gardeners know, this is also very beneficial to many plants. Niche construction need not be a phenomenon limited to the individual or the family group; the niche may be shaped by countless conspecifics present or past. In the case of earthworms, it is not even a particular species, but many different species from the phylum Annelida, that contribute to the environmental modification that enables terrestrial life for these creatures. This activity continues through time and space and stabilizes multiple lineages. But the most striking example of all is the human. We live in hugely elaborate structures within structures, and our ways of life are wholly dependent on this infrastructure. I shall return to this shortly.

Sexual reproduction is a common but not universal feature of evolving lineages. Niche construction is in a sense more nearly universal, as every organism has some effect on its environment. But the construction of complex structures over many generations of the kind exemplified by termites, beavers, or humans is a quite rare accomplishment. All these varieties of lineage stabilization, as opposed to merely external pruning by natural selection, can be seen as stabilization through internal organization, a movement along the spectrum of degrees of internal organization described by

David Hull. The most extreme move along this dimension is the evolution of sociality.

Sociality within a species provides obvious elements of internal organization. These have reached their peak in social insects and in humans, in which a complex division of labour makes possible a system of distinct and interlocking roles. Members of such a species interact with one another in a variety of ways that generally contribute to the maintenance of the complex structure. But social structure also contributes to a rather different feature of a lineage, evolvability. Sociality provides new ways in which lineages can respond adaptively to their environments. To consider these, I must now turn more generally to a more familiar topic in evolutionary theory, the determinants of evolutionary change.

Changing Lineages

Many orthodox evolutionary theorists still assume that the sole or major architect of evolutionary innovation and change is natural selection. I am sceptical of the creative character claimed for natural selection. In one sense, I think it is a simple logical truth that natural selection cannot be creative. Selection is selection. The choices between which the selection is made must come from somewhere else. The counter-argument, perhaps most prominently defended in recent years by Richard Dawkins, is that, provided a very small series of steps can be described leading from the present phenotype to an advantageous future state, small mutations will arise that affect the transition to the next step, and eventually the journey will be completed. Dawkins illustrates this with the case of the vertebrate eye, aiming to rebut the common thought that half an eye is no use to anyone (Dawkins 2006). But why should we believe that a mutation should even be possible to create precisely any envisaged change; and furthermore, bearing in mind that genes typically have many different effects (so-called pleiotropy), and these other effects are very likely to be deleterious, why assume that the envisaged change will be adaptive and therefore selected? We should certainly acknowledge that natural selection, by maintaining stable and integrated lineages, creates the conditions for major evolutionary innovation, but asking it to provide the innovations with no help but small random mutations is very probably asking too much.

Enthusiasm for the overriding importance of natural selection has tended to draw attention away from an essential task in understanding evolution: where do evolutionary changes come from? One possible answer is that these are almost invariably (small) genetic changes or recombinations. But there are many other possible answers, and it would be a serious error to assume in advance that any of them is of predominant importance. So let me mention a few interesting options. I should first remind you that these are changes to the developmental system—that is what the organism is—and may occur at any point in the developing process. As well as (generally small) mutations and recombinations, which are the sources of most interest to orthodox gene-centred evolutionary theory, there are further genetic options. Evolutionary developmental biology (of which more below) has taken great interest in architectural genes in which small changes may produce major morphological effects. It is also possible that horizontal acquisition of genes has had great importance. The mammalian placenta, for example, the defining characteristic of the major group of mammals to which we humans belong, appears to be constructed from tissues that originated from retroviruses inserted into the genome. Epigenetic changes to the genome, changes that affect the expression of genes without changing the sequence of nucleotides, may be directly inherited or may have behavioural effects that lead to their reconstruction from generation to generation. This remains controversial, in part because the inheritance of acquired characteristics has long been something of a taboo in evolutionary thought; but the evidence for inheritance of epigenetic changes is growing steadily (Fitz-James and Cavalli 2022).

Changes to parental behaviour may be learned by offspring. Where a species has developed some kind of social life, there may be transmission of behaviour from non-parental conspecifics. This is the basis of what is known as cultural evolution (see e.g. Mesoudi, Whiten, and Laland 2006). An interesting case here is that of behaviour related to the construction or maintenance of the niche. And this list is by no means exhaustive. The last few suggestions are not available to all species, a point the importance of which I shall elaborate shortly. I might also mention, if only in passing, a fascinating topic in evolutionary theory, the possibility that complexity can arise from undirected drift, so-called constructive neutral evolution (Brunet and Doolittle 2018).

The extension of these sources of variation may also be thought of in relation to the extension of modes of inheritance beyond the familiar genetic. Inheritance methods can be thought of as providing vertical or temporal stabilization of the lineage as opposed to the stability at a time provided by, for example, social or niche structure. But they also provide spaces for evolutionary change. While epigenetic inheritance significantly extends the possibilities allowed by orthodox neo-Darwinism, sociality opens up entirely new pathways. A minimal but very important form of sociality is parental care. This surely qualifies as sociality, in the sense of structured relations between species members with, in this case, distinct roles. A general term for inheritance processes made possible by parental care is 'parental effects'. Parents constantly interact with their developing offspring and these interactions can have important effects on offspring development. This is too familiar from human experience to need much argument, but it is by no means unique to humans.

Cultural inheritance in social species can extend far beyond the parental. In fact, one of its most distinctive features is that transmission of culture can happen between arbitrarily selected species members. Again, this is most obvious in humans, but a wealth of observations on such issues as tool use or foraging choices show that it is a common feature of many other species. There is a long tradition of theoretical work on the dynamics of cultural evolution, but curiously it remains common for evolutionary theorists to insist that it is a phenomenon of only marginal importance, and has little relevance to the theory generally.

This is a good point to contrast what I have been saying about evolving lineages with a common but simplistic understanding of evolution that remains influential. This is the idea deriving from the so-called Modern Synthesis of the mid-twentieth century that evolution is entirely, or almost entirely, a matter of changing frequencies of genes, which have typically small effects on the organism, and which spread through populations by natural selection. I have mentioned this picture in discussing the idea of natural selection as creative, and specifically in relation to the writings of Richard Dawkins. In Dawkins's work, this idea is associated with the suggestion that genes are 'selfish', which, interpreted charitably, means that genes are the drivers of evolution, the things that are selected on the basis of their effects on the essentially passive organism in which they reside.

Much of the dismantling of this rather simplistic story has come about under the rubric of evolutionary developmental biology, or evo-devo (Müller 2007). Evo-devo recognizes that it is the life cycle that evolves and then is well positioned to observe the significance for evolutionary theory of the fact that the determinants of development are complex. This last topic is more strictly associated with the school of Developmental Systems Theory (DST) (Oyama 1985; Griffiths and Gray 1994). While evo-devo has generally focused on the complexity of the embryological processes that underlie development, DST has advocated a more holistic attention to the range of internal and external factors that affect development. Both these movements in evolutionary biology depart from the gene-centred orthodoxy by recognizing the fundamental role of the individual organism in evolution. Moreover, they recognize that the organism itself must be seen as a developmental system. Both, in different ways, have looked at the developing system as the fundamental unit that evolution has produced. They are, therefore, if not always explicitly, process theories.[4] While I am generally entirely sympathetic to these perspectives, in this chapter I have focused not on the individual developmental system, but on the larger and longer-lasting entities composed by these developmental systems, lineages, or species. In the final part of this chapter I shall make some rather different remarks about these remarkable phenomena.

Species and Lineages

Let me return to the two senses of 'species'. In the classificatory sense, every organism belongs to a species. If we find an organism that we cannot assign to an existing species name, we invent a new one. Every organism belongs to exactly one species. But the species as a material entity, a continuant process, is something quite different. I said at the beginning of this chapter that the lineage or species was what evolved. But it is not merely, as I stated then, that the distribution of properties among its members changes as it evolves. The nature of the process also changes, and the highly integrated species that occur in social animals are themselves highly evolved entities. Not all species in the taxonomic sense are integrated or stable lineages.

As I mentioned earlier, it is questionable whether bacteria—or archaea, the other great group of nucleus-lacking microbes—belong to integrated lineages at all. (Henceforward, when I say 'bacteria' I will take this to include archaea.) The lineage of a bacterium may simply be the sequence of its ancestral cells, and classification of bacteria may be based on quite different factors. It is likely that natural selection will maintain loose populations of similar cells, but these may be of brief duration. Bacterial taxonomy was originally grounded in pathological properties, though this covers a minute proportion of the real diversity. More recently, various measures of genomic similarity have been used. It has also been suggested that most bacteria do exist in something surprisingly similar to the sex-based species, by virtue of

---

[4] The processual character of DST is explicitly articulated by Griffiths and Stotz (2018). Robert, Hall, and Olson (2001) explore the relations between DST and evo-devo.

exchange of genetic material within a restricted group (Bobay and Ochman 2017). But no one claims that this is true of all bacteria and even if it were, there is no reason to assume that these groups would provide a universally optimal ground for classification.[5] More to present purposes, the sociality, parental relations, and so on that I have been discussing do not seem to apply to such species, though in the next chapter I shall discuss respects in which bacteria might seem highly social.

In the case of multicellular organisms, we can note an important distinction, if ultimately a matter of degree rather than kind, between what have sometimes been known as r-selected and K-selected species. r-selection is the production of sometimes enormous numbers of offspring and leaving them entirely to fend for themselves. The prize winner in this respect is said to be the Common Mola or Ocean Sunfish, which lays about 300 million eggs. The more familiar cod can produce about 5 million. Many insects also produce very large numbers of eggs. Obviously, the overwhelming majority of these fail to reach maturity. But they do provide the opportunity for a lot of natural selection. In such species there is typically little integration of the species beyond natural selection and mechanisms for maintaining reproductive isolation. Among vertebrates, amphibians such as frogs may lay thousands of eggs, and reptiles may lay dozens. Again there is usually little parental attention.

K-selection is the move to high investment in a small number of offspring. This reaches its peak with birds and many mammals, invariably exhibiting a good deal of parental care, as mandated by breast-feeding or nest-feeding, and often a wider social structure. The lineage in these cases is not only more strongly integrated, but also provides new evolutionary possibilities by facilitating a range of parental effects and cultural transmission. Local concentration and cooperation also make possible the most impressive kinds of niche construction, such as exhibited by beavers and humans or, in a rather different evolutionary trajectory, the social insects.

Having mentioned cooperation, I should make a very important general point here. Evolutionists often like to talk about competition. And no doubt this has been a vital driver of evolution. But as evolution progresses beyond merely the changing properties of the individuals that compose a lineage,

---

[5] While concordance with phylogenetic history is still generally considered a desideratum of bacterial taxonomy, for reasons discussed in the text, especially the prevalence of lateral gene transfer, it is unclear whether this is even in principle an achievable goal. In practice, increasing access to genomic data has favoured classification solely on the basis of genomic similarity. See e.g. Hugenholtz et al. 2021.

to changing properties of the lineage itself, cooperation becomes an increasingly important element of the evolutionary process. Cooperation enables the transmission of information between unrelated individuals, thus fostering far more effective cultural evolution, it may allow more effective care of offspring through recruitment of helpers, and it allows far more elaborate transgenerational ventures in niche construction. No doubt, as traditional advocates of the overwhelming predominance of competition will object, there will be some conflict between selfish individuals and the interests of a cooperative society. It is even argued that this individual competition will inevitably undermine cooperative ventures. But the presence of highly cooperative species shows that this conflict can be overcome. No doubt one reason for this is that cooperation can be very effective in identifying, isolating, and sanctioning individuals who attempt to advance their own interests over that of the social whole.

The success of cooperative lineages is sufficiently demonstrated by the extraordinary successes of social insects and humans. Humans and their livestock make up over 95 per cent of the vertebrate animals on the planet. And for every human on the planet there are said to be 2.5 million ants—much smaller than us, but a lot of ants. These groups point to a final and perhaps decisive benefit of cooperation, the enabling of a division of labour.

This leads me to conclude the present chapter with a few words on the human lineage, which will be my primary focus for the remainder of this book. And here I shall shamelessly espouse human exceptionalism, not perhaps as a matter of kind, but certainly as a very large difference in degree. The human lineage is unique in the degree of its integration and in the complexity of the infrastructural niche it has constructed. A huge part of what has made this possible is the extreme and flexible division of labour that is unique to humans. And, a topic to which I shall return in chapter 5, what makes this division of labour possible is the unique developmental plasticity of humans with respect to their behavioural capacities.

Just reflect for a moment on the variety of people whose efforts jointly made possible the public event at which this chapter was first presented. Some that come to mind are architects, builders, sound and lighting engineers, IT specialists, clothes makers, the designers, manufacturers, and operators of the aircraft on which I arrived in Edinburgh, likewise for the taxi in which I came from the airport, the hotel in which I was accommodated, and on and on. These roles constantly change, moreover, as typewriter makers and switchboard operators cease to exist and are replaced by bloggers and computer programmers. Of greatest relevance to our present concern, this has created a situation in which genetic bases for evolution have become far less important than social and culturally mediated processes of change. (And I don't mean to imply that the former may not also remain important.) The schools, television programmes, books, universities, and so on that shape the development of humans today are very different from those that did so a few decades ago. Now, for better or worse, we even have professional influencers. And the technologies that have emerged over the last two centuries have changed profoundly who we are. The mobile phones that have existed only for a quarter of a century are now as standard a piece of human equipment as shoes or wallets, and far more so than watches. I would argue that by making friends and family instantly accessible at any time and in any place, it has wholly transformed our experience of social space, which no longer coincides with geographical space.

The dominant position of humans on our planet, I suggest, is as much the result of the evolution of the lineage itself as of the individual organisms of which it is composed. It is ironic in the light of this that an ideal of autonomous individuality has been so powerful in recent centuries. For there is surely no doubt that it is the evolved capacities of our species as such, not of individual organisms, that is our peculiar strength. I shall return to related thoughts at several points in later chapters.

But for now, let me take stock. Over these first two chapters I have discussed two of the most important kinds of biological system, the organism and the lineage. Without denying the enormous interest of lower levels of organization, such as the cellular, molecular, or genomic, I believe that these enable rather than determine the activities of the more complex systems I have discussed, and so they do not require detailed attention for my present concerns. Organisms and lineages, I have argued, are persistent processes, an ontological category that has been insufficiently stressed by philosophers. They are also highly interdependent processes. Lineages are composed of organisms and persist because organisms reproduce. But organisms are generated by lineages, and have the properties they do because of the particular evolution of the lineages to which they belong.

A feature of processes that I have occasionally emphasized is their lack of sharp boundaries and the consequent possibility of their being deeply and even inextricably intertwined. In the next chapter I shall pursue this theme with regard to the entanglement of organisms: first in connection with the ubiquity of symbiosis in the living world, and second with respect to the gradual disentanglement and divergence of processes in reproduction.

3

Humans and Their Fellow Travellers

Organisms

In chapter 1 I argued that organisms were a kind of process. But this argument largely took it for granted that we know what we are talking about; we know an organism when we see it. I begin this chapter with a rather different approach to the question, What is an organism?

Here is a standard view. There are two kinds of organisms, single-celled and multicellular. A single-celled organism, unsurprisingly, is a single cell. Its boundary is the outside of the membrane that surrounds the cell or if, as is the case with bacteria, algae, fungi, and some protists, there is a partially rigid cell wall outside the cell membrane, then that is the boundary. Multicellular organisms originate with a founding zygote which then continues to divide and differentiate. The sum of all the interconnected cells derived from the zygote constitutes the organism. The outside layer of cells, or those bits exposed to the outside world, form the surface of the organism. This will consist of either cell membranes or cell walls. Or so the story goes.

However, a number of biologists and philosophers have recently insisted on a problem with this story. Starting with single-celled organisms, although there are many cells for which this story works well enough, the great majority of bacteria and archaea are found in complex communities, either of their own kind or, very often, containing many different varieties of microbes. Especially interesting among these are the so-called *biofilms* that form on almost any damp surface, from the slimy surfaces of rocks in a stream, to the plaque that forms on our teeth (Donlan 2002).[1] They have been called the most successful life form on Earth (Flemming and Wingender 2010). These communities have a characteristic life cycle, recruiting various kinds of member in a particular order and eventually dispersing. The bacteria in a

---

[1] Biofilms are of enormous human significance, as a challenge e.g. to water supply systems and human health, as an essential component of e.g. healthy soil and healthy guts, and as a potential resource for many industrial uses such as chemical engineering. Their importance would be hard to exaggerate.

Everyone Flows. John Dupre, Oxford University Press. © John Dupré 2025.
DOI: 10.1093/9780198941866.003.0003

biofilm excrete compounds that form the so-called extracellular matrix. This keeps the residents in a more or less fixed location and protects them from a range of threats. It also allows the recycling of waste products from dead cells and even stores genetic material for potential lateral transfer. And biofilms exhibit a division of labour analogous to that between the various tissues and organs of a multicellular organism. Some cells adhere to the surface on which the biofilm forms, others provide sequential chemical stages in the metabolism of the system's food. Still others specialize in generating the constituents of the extracellular matrix.

Given all this, there is an excellent case for considering biofilms to be a kind of organism. They are functional wholes, able to provide the means for their persistence over time, and exhibiting a characteristic life cycle. Many would still insist that they are better seen as ecological communities in which various individual organisms (bacterial cells) come together and interact for their mutual benefit. But the question then arises, if the constituent cells of the biofilms cannot function fully in isolation, should *they* be seen as organisms? It is true that these cells can survive in isolation for a time, as they do when dispersed from the biofilm. But if they cannot grow or reproduce until they find a fresh community or biofilm in which to insert themselves, they should perhaps be seen as more like free floating gametes than complete organisms. This is not a simple issue to resolve, but it does indeed seem that many of the cells involved in biocells are unable to grow and divide outside the peculiar context of a biofilm.

Biological orthodoxy generally rejects the suggestion that a biofilm is an organism, for two related kinds of reason. First, evolutionary theory does not predict cooperation between different kinds of organisms. The different constituents of the biofilm are very likely to be genetically homogeneous clones, which are indeed expected to cooperate. But clones of distinct kinds should be seen rather as competitors for space and resources than as cooperators. As is often the case, the balance between cooperation and competition is a delicate one, and the situation within a biofilm is by no means exclusively cooperative. Much of the stability of the system is maintained by the proper spatial arrangement of distinct clones. Also sophisticated sensory mechanisms such as quorum sensing help to minimize antagonistic interactions (Nadell, Drescher, and Foster 2016). Evidently, given the prevalence of biofilms, these processes are effective.

Second, related to the first problem, biofilms are not generally seen as forming identifiable lineages. When the component cells in the biofilm disperse they may end up rejoining a significantly different set of cells to form a different biofilm. The lineages in which evolutionary change can occur remain the lineages formed by the constituent cell rather than any lineage of biofilms. One might, nonetheless, ask why the role in evolutionary theory should be more important than the ability to survive and reproduce in determining the proper reference of the term 'organism'. A biofilm is, at least, a coherently organized biological individual.[2] Moreover, as already noted, it is not clear that many of the single-celled constituents of biofilms are able to function independently.

This problem is strikingly reflected in the current state of bacterial taxonomy. Traditional rules of biological nomenclature require that a material instance of a named organism be held in some suitable collection or museum. For bacteria, the required material is a cell culture. But the vast majority of bacteria cannot be cultured in a laboratory, and in many cases it is likely that this is a consequence of the fact that they are obligately social members of multispecies communities. While there are estimated to be millions of bacterial species, only about 20,000 have been named (Hugenholtz et al. 2021). Something seems to have gone wrong.

If the correct diagnosis is that the biofilm is actually a better candidate for being an organism than the individual bacterial cell, this has profound implications also for how we should think about the multicellular organism. As I mentioned in chapter 1, multicellular organisms, in the standard sense that includes only the cells derived directly from the founding zygote, are not, at least in the vast majority of cases, viable in isolation and also require symbiotic relations with a great many other types of cell. Considering just the human body, somewhere between 50 and 80 per cent of the cells in it are non-human. Fellow travellers are mainly bacteria, but also include archaea, protists, and fungi. There are also, as I shall discuss shortly, countless viruses.

Not all these non-human cells are beneficial. Some may be seriously pathological, and some may be neutral, doing little harm or good. Interestingly, whether a kind of cell is beneficial or pathological may depend very much on circumstances, especially on its location (Méthot and Alizon 2014). Cells that are beneficial in the gut may be lethally toxic in other parts of the body.

[2] The question of biological individuality has been widely discussed in recent philosophy of biology. Much of this discussion, especially in relation to evolution, starts from the account of Godfrey-Smith (2009). Papers addressing the case of biofilms include Ereshefsky and Pedroso 2013, Clarke 2016, and Dupré and O'Malley 2009. I shall say more about this topic in the next chapter. I shall not try here to distinguish sharply between an organism and a biological individual. Certainly the latter term is broader, though neither is simple to define, and the former is a subset of the latter. One aim of this chapter is to show the difficulties in dividing the living world into individuals of any kind and, a fortiori, of dividing it into organisms.

What is important and now widely agreed is that many of these bacteria are essential for the health of the human. Digestion, development, and immune function are only the most clearly established functions in which symbionts play essential roles. It is clear, in short, that the healthy human is a multi-species symbiosis.

Holobionts

It has become common to refer to a multicellular organism together with all its symbiotic partners as a holobiont. (Hereafter I shall sometimes refer to a multicellular organism excluding symbionts as a 'macrobe' (Dupré and O'Malley 2009).) And it is increasingly common among philosophers and some biologists to argue that it is the holobiont that best qualifies as the human individual. I am generally sympathetic to this proposal, but there is a problem, exactly parallel to that described for the microbial biofilm. Holobionts don't form lineages of the kind that I discussed in the last chapter. Some symbionts are passed on by parents, and can be treated as a parallel mode of inheritance fully aligned with genetic transmission. Here we might mention the mitochondria, internal constituents of the eukaryote cell, often referred to as the powerplants of the cell. It is widely accepted that two billion years ago the ancestors of the mitochondria were free-living bacteria, subsequently enveloped by another early cell. After two billion years of evolution, mitochondria have come to specialize in a few vital metabolic functions and have lost many of the functions they would require to live independently. At this point there is no theoretical reason to think of the mitochondrion as a distinct entity.

But many symbionts are recruited from the environment. And indeed, it is highly desirable that this should be the case, because the recruitment of local bacteria is likely to be an important way of tracking and responding to changes in the environment.[3] But in that case, the lineage of the bacterial symbiont is quite distinct from that of the host macrobe, and hence holobionts are not lineage forming; and if they are not lineage forming, it appears that they are not the individuals that evolved. This problem is, it seems

---

[3] Genes from ingested bacteria in the environment may also be transferred to resident bacteria, as appears to be the case for genes that enable the digestion of the seaweed used to wrap sushi (nori), found in many Japanese gut bacteria, but seldom in European or American microbiomes (Hehemann et al. 2010).

to me, unanswerable in a world of things. If a human is a thing that evolved, but the thing that evolved cannot function in the way a human does, we have an impasse.

But recognizing that we live in a world of process offers a straightforward way out. Let us agree that what has evolved is what is described by the orthodox concept of an organism: either a single cell, or a cell lineage. I have sometimes referred to the latter, as the set of cells derived from a macrobial zygote, as a monogenomic differentiated cell lineage (MDCL), to point to the facts that the cells in such a lineage share (more or less) the same genome, but include many different phenotypes (brain cells, muscle cells, liver cells, etc.), and I shall use this term in what follows (Dupré 2011).[4] Since the sperm and egg that formed the initiating zygote both came from members of the humans species, an evolving lineage, we can trace this backwards down the human lineage. So humans qua MDCLs are part of an evolving process traceable back through countless ancestral species to LUCA, the last universal common ancestor. This is the perspective that makes a biologist focused on evolution think of the human organism as the MDCL.

Now consider a bacterium that is an obligate symbiont of the human. This is also part of an evolving process. This process and the human MDCL lineage are perfectly intertwined. Every component of the bacterial lineage is embedded in the human lineage. Yet the two differ completely in the ways the processes persist; the one by cell division, the other by sexual reproduction and a long, complex developmental process. Nonetheless, the two lineages help to stabilize each other. The bacteria are entirely dependent on the humans to provide their living environment, and the bacteria, *ex hypothesi*, provide some necessary service to the humans. Now contrast an environmentally acquired but essential symbiont. Very likely this microbe does not require the specifically human environment; it was just hanging around waiting for somewhere to go. Perhaps it can live in other species or even in freestanding biofilms. The lineage of this organism bumps into the human lineage here and there, but is not constantly intertwined with it.

Finally, consider a human holobiont. Central to it is an element of the human lineage. This inevitably comes with a certain number of microbial symbionts with which the human lineage is fully intertwined. Many other

---

[4] I should confess, as will be evident from the preceding chapter, that this is something of a misnomer. The set of cells derived from the founding zygote is really a clade, not a lineage. The lineage leading to my liver cells is not the same as that leading to my brain cells, though both converge at the zygote, the universal common ancestor. But I will not worry about this here.

microbes will pass through it, some essential contributors to the system drawn from the environment, others more or less neutral or harmful in their contributions. A considerable part of this microbial flow through the system contributes to the stabilization of the whole, though some parts, pathogens, may threaten its destabilization. Is there any answer to exactly how much of all this is truly part of the functional whole? Surely not. But this is just what processes are often like. As I explained in chapter 1, processes do not have clearly defined boundaries. This is because, as in the present case, they are stabilized by a host of interactions at their boundaries. Think once more of an eddy or a wave. There is no place where the eddy ends and the surrounding stream begins. The flow of energy from the stream to the maintenance of the pattern that is the eddy is a continuous process with no defined border.

Quite generally, processes can interact and intertwine while stabilizing (or destabilizing) one another. Think, for instance, of social or economic processes, such as unemployment, inflation, or the welfare system. Individuals engage with these larger processes. For example, becoming unemployed may reduce their expenditure, thus contributing to a downward pressure on inflation and increasing the flow of resources from the welfare system. Individual economic acts cannot be uniquely partitioned between such processes and all of them aggregate to form flows such as national income. What is included in these larger aggregates, finally, is a matter of decision, though sometimes one with considerable normative if not metaphysical importance. Think, for example, of the long-standing debate over whether unpaid domestic work should count as part of GDP (Waring 1988), a decision which has decisive effects on the well-being of many people, mainly women.

Much the same is true of the holobiont. Living material flows through this complex system and, depending on its importance and its length of stay, may have a greater or lesser claim to be included as a constituent of the persistent whole. Thinking of a human as a holobiont, we may worry whether the surface of the living system is the human skin or the complex communities of microbes that cover it. The answer, I suggest, is that either may be correct, and the decisive question is why one is asking the factor. In the context of central evolutionary questions, it may be appropriate to think of the human as the MDCL. For many physiological questions it will be necessary to think of a more expansive holobiont.

And this looseness of boundary, the possibility of distinguishing different, overlapping systems within the flow of biological material, is not an objection but a virtue for a process ontology; not a bug but a feature. This becomes clear when we explore the relation of the individual human to larger systems of which it is part. But before discussing this topic, I want to look at one further, rather surprising component of the holobiont, the virome.

Viruses

I have mentioned that there may be several times as many bacteria in the human body as human cells—though recent estimates have suggested the numbers may be more nearly equal. As this discrepancy suggests, this is not an easy thing to measure. But for present purposes the fact that there is a very large number, certainly in the tens of trillions—which is a lot—will suffice. There are also a great many viruses, probably about ten times as many as there are bacteria. Just as it was for a long time thought that bacteria were at best harmless and frequently dangerous, but recently it has been realized that many are highly beneficial or even essential for the lives of the multicellular organisms that they inhabit, so it is with viruses. While even many scientists continue to see them as generally pathological, it is gradually becoming clear that many, perhaps most, are harmless or beneficial.

The great majority of the viruses living in or on a human are phages, viruses that reproduce in bacterial cells rather than human cells. Natural hypotheses are that these viruses might be functional for the holobiont by protecting against pathogenic bacteria or, more interestingly, regulating the numbers of symbiotic bacteria. Evidence for the first function is provided by the high concentration of phages on the mucus tissues of the mouth and other orifices, obvious places for the potentially dangerous encounter with harmful bacteria.

The second function is supported by a rather more theoretical argument. Presumably the maintenance of these large numbers of viruses implies that their hosts are the symbiotic bacteria that live in the human body, suggesting that they are predators on these generally friendly bacteria. But mathematical ecologists have long studied the cyclical oscillations characteristic of populations of predator and prey organisms. The battle between a virus and its bacterial host would be expected to follow a similar pattern, as large reductions in the bacterial population due to viral predation eventually bring it to a level where viruses cannot find prey, and their population collapses in turn, allowing the bacterial populations to regenerate. The lack of any such dynamics is suggestive of the absence of such straightforwardly predatory interaction. It seems, in fact, that the population of viruses in a human is fairly stable, though subject to disease-specific changes (Shkoporov et al.

2019). Waves of a virus may infest a host as a pathological infection, but the background virome, as the whole viral population is known, remains much the same. This suggests that the virome is part of a stabilizing system that, perhaps in conjunction with the human immune system, helps to maintain stable and optimal populations of bacteria.

There are also more subtle ways in which viruses help to combat microbial disease. Several harmless viruses have been shown to protect against pathological viruses. For example, human pegivirus (HPgV-1) appears to have a beneficial effect on a number of harmful viral infections, including HIV (Yu et al. 2022). Examples of benefits supplied by viruses to their multicellular hosts are growing all the time (Roossinck 2011). We should, perhaps, find this quite unsurprising. The image of the living world provided by the holobiont, in which multiple different organisms or, better, lineage segments of organisms, since the time scales of the kinds involved in these complex systems are very different, come together to provide stabilities in the living flux, makes it natural to ask of the 10^{13} virus particles that exist in our bodies (Liang and Bushman 2021), not How do we survive their hostile intentions?, but, What are they contributing to the wholes of which we are part?

How is the enormously complex system maintained, or stabilized? One very important way is through the immune system. Contrary to the still widely disseminated self/other theory, according to which the immune system simply distinguishes human from non-human cells and attacks the latter, insights into symbiosis have been one factor that has led to a more sophisticated view (Pradeu 2011). Clearly the immune system must somehow distinguish desirable from pathogenic microbes, both cellular and viral. And it is also very active in disposal of dead or diseased human cells. Given the fact that the difference between beneficial and pathogenic microbes can be place dependent, this is a difficult task. One suggestion is that the immune system does not respond in the first instance to the intrinsic nature of the entity with which it interacts, but rather to sudden changes in the chemical environment.[5] If that is correct it is rather literally a stabilizing system.

Second, interbacterial interactions can maintain the stability of the bacterial communities that are part of a holobiont, and thus the holobiont itself. Much of this is achieved through the cooperative construction of biofilms, in which the extracellular matrix protects the community from many

---

[5] Such a view has been strongly championed by Thomas Pradeu (2011, 2019).

threats, including antimicrobials. (Hence the ability of humans to take anti-microbial drugs without suffering catastrophic systemic damage.) As already noted, it need not be assumed that interactions within a biofilm are solely cooperative, but various processes such as quorum sensing (Miller and Bassler 2001) help to control antagonistic interactions dangerous to the stability of the biofilm (Nadell, Drescher, and Foster 2016). More complex systems involving host/bacteria/phage interactions may also be fundamental to system stabilization.

In sum, there are very strong reasons not to think of the human individual as merely composed of what have traditionally been thought of as human cells, but as also including a multitude of bacteria, viruses, and perhaps protists, fungi, archaea, and more. But there still may seem to be a problem for how these complex individuals, holobionts, evolve. If they are, indeed, the functional individuals that forage, eat, mate, and so on, they are presumably the individuals that are exposed to natural selection; they are, as it is said, the units of selection. But if the units of selection are not capable of reproducing, then selection will not have any effect on the distribution of traits in the next generation. I have already said that for (some) evolutionary purposes we should still consider the MDCL as an individual. But I want now to look more critically at the argument that the holobiont cannot function perfectly well as such a unit and even qualify as lineage forming.

Holobiont Lineages

First, we should again note the difference between symbionts that are passed on by the parent and those that are recruited from the environment. In the former case we have simply an alternative path of inheritance from the parent to the offspring, and so no problem. A well-studied case is the many insects that carry bacteria inside their cells, bacteria that have long ago lost the ability to function independently. In some cases it is not clear that the embedded bacteria, such as *Wolbachia*, are wholly beneficial to the insect, but this doesn't matter. The status of the bacterium in evolution is no different from that of the nucleus, also transferring a packet of genes between generations. With luck, over time, natural selection will better align the interests of the insect with the effects of its captive bacterium.

Environmentally recruited bacteria may seem more problematic. If only a small proportion of the bacteria live in human holobionts, then a human whose success is augmented by a beneficial change to the bacterium will have little effect on the likelihood that future generations of humans will recruit the improved bacterium. However, there are several caveats. First, if the bacteria are generally symbionts of mammals, say, a benefit to one mammal will often be a benefit to another. Second, recruitment is unlikely to be random. Holobionts constantly shed symbiotic microbes, and off-spring are therefore generally more likely to recruit symbionts that originated from their parents, simply because these are in the vicinity. Moreover, there are various ways by which symbionts are passed on to offspring that are surely evolved. Mothers infect offspring with bacteria from the birth canal, and the niche may well be constructed in ways that exclude inappropriate potential partners.

But finally, not every aspect of the holobiont needs to be equally evolvable. This is true even of MDCLs as units of selection. Different parts of the genome are subject to variable mutation rates, for instance. And even if none of the factors just mentioned applies, a successful mutation in a symbiont will, however minimally, increase the probability of finding this mutation in the next generation. The upshot of this, I suggest, is that there is no reason to deny that holobionts are fully lineage-forming entities. Where the population of lineages is bounded by relations of reproductive connection, that is, the large host organism forms a well-defined species, there is no reason not to describe this as a lineage of holobionts. This clearly applies to humans. There is an interesting question whether this argument can be made for symbiotic systems that lack the boundary-providing large multicellular organism, for example for biofilms. I am not convinced that there are not lineages of biofilms, but this is an argument for another time.

As is a theme of this book, for any living process we can ask what it is that sustains it in a condition that is always far from thermal equilibrium, always under threat from the force of entropy. Life, as I conceive it, is quite generally an intertwining of parts of lineage processes to produce stable individuals (Dupré and O'Malley 2009). These individuals, in turn, generate new parts—organisms—to sustain the lineages that they include. There are several aspects to the stabilization of these individuals. There are the internal reactions generally considered as physiological, and I think that important symbiotic processes naturally belong within the science of physiology. There is a wide range of ecological relations with other lineages, as predators and prey or just modulators of the environment, such as the earthworms that increase the fertility of the soil and the availability of plant food. At the extreme, are highly evolved mutualisms. A famous example is the relationship between ants and acacia trees, in which the ants protect the trees from a wide range of predators, while the tree provides the ants with shelter and food.[6] The relations between humans and their various livestock species might also be considered mutualistic. The large majority of birds on the planet are chickens and ducks raised by humans. Whether this technical evolutionary success provides any adequate justification for the often appalling conditions in which these animals live is another question. Ants also take care of livestock, such as aphids and scale insects, which they protect from predators and milk for their honeydew. There will be a few more points in this book at which I will remark the interesting parallels between humans and social insects.

Of even greater importance for humans is the way their lives are sustained by their relations to other humans. But before turning to that topic more generally, I want to say a little about another aspect of human individuation that I think can be much illuminated by a process ontology, pregnancy.

Pregnancy

A common understanding of pregnancy is that a new life begins when an egg is fertilized. From the moment of fertilization, a woman is both an organism and a container for another organism. This is the 'bun in the oven' view of pregnancy. The mother is cooking, or anyhow incubating, the new individual. One might suppose that considerations of symbiosis show that we are already known to be hosts to countless other organisms. But physiologists tend to think of the human as a hollow tube, with the alimentary canal strictly on the outside. On the holobiont view, several pounds of bacteria within the tube are counted as part of the human organism. But whether one thinks that the symbionts are separate organisms external to the body or parts of the body, they are not distinct organisms within the body. So the container view of pregnancy proposes an unusual, perhaps unique, ontological situation.

In parallel to the proposal we have been considering for microbial symbionts, might we rather take the foetus to be a part of the mother? It is interesting to compare the many representations of the pregnant female that

---

[6] This well-known example is actually much more complicated than is often noticed, involving several species of ants with differing preference for tree sizes, offering more or less benefit for the host, and interacting with one another in more or less antagonistic ways (Young, Stubblefield, and Isbell 1996).

emphasize the distinctiveness of mother and foetus, with ultrasound photographic images in which the distinction is often much less sharp, anyhow not much more distinct than an image of a heart or a liver. The proposal that the foetus is better seen as a part of the mother has been strongly advocated by Elselijn Kingma (2018, 2019), whose work has also done much to bring the question to the attention of philosophers, and the proposal has much to be said for it. The relations between a foetus and a pregnant female are intimate beyond any other between distinct organisms. All the flows of nutrients, oxygen, and other chemicals essential for its well-being pass through the mother to the foetus. It certainly does not qualify as an autonomous thing or substance. It is true that it isn't exactly autonomous even when the umbilical cord has been cut, but it does at least have its own reasonably clear boundaries.

What makes this debate so important is, of course, its relevance to the question when a human life begins, and hence to the ethics of abortion. The answer that it begins at fertilization is philosophically coherent—a process that may eventually lead to a human life begins there—but the enormous difference between a single cell and a developed human being makes the answer very unhelpful. Those who believe that the right to life begins at fertilization generally base their claim not on the nature and status of this particular human cell but, if not on a religious corollary to this, for example that it is the moment of ensoulment, at least on a philosophical claim, that a new essence has appeared. I do not believe either of these claims; but neither is anyhow well suited to the defence of a right to life. Religious beliefs are far too diverse to ground public policy. No one whose religion forbids it should be required to have an abortion; but nor should the beliefs of the religious be imposed on those with different or no religion. Essentialism is a fairly technical philosophical doctrine, and is probably not deployed much in this debate. But, or so I have argued extensively over the years and in this book, it is anyhow simply false: biological systems don't have essences (Dupré 1993).

The philosophical problem that emerges if we recognize the inadequacy of conception for this purpose, is that of finding any well-motivated answer to the question when the developing foetus should be counted as a person with rights. In the debate over the ethics of abortion many sensible people agree that foetuses past a certain developmental stage deserve at least respect, whereas similar claims for fertilized eggs are much harder to sustain. Most people are not greatly concerned by the numbers of such cells stored in liquid nitrogen, some of which may be transplanted into women and most of which will eventually be disposed of. So where should we draw the line?

One line has been drawn by a number of ethicists, though for a different purpose, defining the cut-off point before which it is permissible to experiment on human embryos. This is the appearance of the primitive streak, an important developmental structure that appears about fourteen days after fertilization. The widespread agreement on the significance of this cut-off point dates from the influential Warnock Report on human fertilization and embryology (Warnock 1984), which recommended that 'no live human embryo derived from in vitro fertilisation, whether frozen or unfrozen, may be kept alive, if not transferred to a woman, beyond fourteen days after fertilisation'. Interestingly, the report is quite clear that human development is a continuous process and any such boundary is ultimately arbitrary. But for practical, specifically legal and regulatory, purposes, a clear line needed to be drawn. They decided on this particular point on the basis of a number of recommendations from a variety of expert bodies appealing to criteria such as implantation and the earliest stages of neural development.

However, this has not prevented philosophers from offering justifications of the significance of this particular stage of development. The issue of greatest interest to philosophers has been the possibility of twinning, which also ends at about the fourteen-day mark. Here is the problem that arises from possible twinning. Suppose you are the same individual as your week-old embryo. But that embryo might have twinned and been identical to another person. Then, your twin would also have been identical to the same embryo and, since identity is transitive (if a is identical to b and b is identical to c, a is identical to c) you would have been identical to your twin. But that is clearly impossible, and the demonstration that it is possible provides a reductio ad absurdum of the original premise. So, it is concluded, the person you are cannot have come into existence until twinning is no longer possible.[7]

If this argument strikes you as fishy, so it should. The possibility of the embryo turning into two people rather than one should be an argument against

---

[7] More or less nuanced endorsements of this argument can be found by e.g. Eric Olson (1997) and David Wiggins (e.g. 2016). Much of its significance turns on the question whether an embryo is a human being, and the conviction that it is morally problematic to experiment on human beings, at least without the consent that embryos are unable to offer. Typically, the argument is taken to show that the human being comes into existence at fourteen days. Oderberg (1997) argues that twinning is actually to be seen as an unusual way for a human being to come to an end. As one might say with a fissioning microbe, one cell ceases to exist and another two come into existence. In most cases, the human being comes into existence at the moment of fertilization. In less common cases two human beings come into existence at the point of twinning. The embryo destined to twin is a human being with all the moral status that that may imply, but a very short-lived one.

experimenting with it, not the opposite. And returning to the question of abortion, most advocates of women's right to choose would put the stage at which a pregnancy can be terminated a great deal later than two weeks.

Kingma's proposal that the embryo and then the foetus are parts of the woman's body fits well with the common formulation of abortion rights in terms of a woman's control over her own body. Abortion, from this point of view, amounts to elective amputation of a non-essential part of the body. But the view does face a serious dilemma. On one horn of the dilemma, the new individual does not come into being until birth. Prior to that there is one in-dividual, afterwards two. This does put a very heavy philosophical burden on the act of birth. This is, of course, a far from arbitrary event in the life history of mother and baby, but whether it is sufficiently so to constitute a coming into being of an individual *de novo* is another matter.

On the other horn of the dilemma, we allow that the new individual pre-ceded its birth, but also that it did so as part of another individual, its mother. As Kingma discusses, this view explicitly acknowledges that one organism can be part of another organism, which she notes to be a problematic conces-sion within the substance ontology to which she still adheres. She is, there-fore, reluctant to embrace either horn of the dilemma.

Personally, I am happy to embrace the second horn of the dilemma. But I do so because in a process ontology such a part/whole relation just isn't a problem. Kingma, however, explicitly accepts a substance ontology, and per-haps for this reason seems unwilling to endorse either horn of the dilemma. For, as I noted in chapter 1, one of the most characteristic features of a sub-stance or thing ontology is the autonomy of a thing. And it is hard to be less autonomous than being part of something else.

How should we describe pregnancy within a process ontology?[8] Easily, I think. If life is a flow of activity, then pregnancy is a bifurcation within that flow. Bifurcation is itself a process that takes a considerable time. Indeed, though birth is a highly significant event, it is hardly a point at which bifurca-tion is fully achieved and the two processes are fully separated. It is true that it does become possible, if often traumatic, to separate the processes at the point of birth. But this point only shows that it is not possible at any earlier stage in the most trivial way. Birth just *is* the separation of mother and foetus, subsequently reclassified as infant. Compare here the gradual bifurcation of a species as a subpopulation becomes gradually isolated from the parent

---

[8] A detailed answer to this question is provided by Meincke (2022).

population, though reproductive links continue as they diminish in frequency. There is no precise moment at which separation has been achieved even though there comes a time in the future when there is no doubt that this has happened.

I have not answered the question when a developing human process becomes a person with ethical rights; indeed, I have refused to do so. There are, I think, only two points within the biological reality that provide any clear basis for such a decision, fertilization and birth. But for reasons I have indicated there is no widely acceptable basis for marking a line at fertilization; and few people will wish to defend the acceptability of fortieth-week abortions on the ground of an absolute ontological divide between the foetus and the neonate.

As I shall discuss in much more detail in the next chapter, the beginning and end of a process, and specifically a human process, are to a substantial degree conventional decisions, properly based on pragmatic considerations. Compare the convention that in some cultures the eighteenth birthday marks the divide between adolescent and adult: some such boundary is necessary, and that seems (to us, now) about the right time. Moreover, separating a process from its surroundings, or from a process with which it is intertwined, is not something it does for us, but something we do conceptually. It is unquestionably natural to differentiate the foetus conceptually from the mother. It has a relatively sharp boundary, it has a distinctive genetic structure, and it has a unique expected fate. But, on the other hand, it is intimately connected to the mother by material and energetic flows, and by mutual influences on the physiology of both parties. It is, in short, a distinct and differentiating process that is part of a larger process. It will not be easy to decide what stages in the development of this process mandate or preclude particular kinds of treatment given the gradually divergent interests of the two separating parts of the process. But at least we can do so on the basis of discussion of relevant features such as sensibility or consciousness, and with proper respect to both parts of the process that we wish to separate. To ground such a discussion in hypotheticals about possible twins serves only to show the deficiencies of a substance ontology.

Human Cooperation

Any organism is a process sustained by internal and external processes. It is also a part of a hierarchy of processes—molecule, cell, organism, lineage—which, to varying degrees sustain one another. Most lineages generate organisms without doing much subsequently to stabilize them. But as I discussed in the last chapter, the evolution of new characteristics of the lineage itself has eventually led to the emergence of social interactions that do play a major part in the sustenance of the species' members. Humans, as I then noted, are unique in the way they have developed social interactions and possibilities, though perhaps the eusociality of the Hymenoptera (ants, wasps, and bees) and termites is a comparably successful alternative route.

The understanding of life as involving the mutual stabilization of processes already shows, in opposition to the extreme emphasis on competition characteristic of much evolutionary thought, that life is a highly cooperative process. Nonetheless, the uniqueness of human cooperation is hard to overstate. It is the ability to deploy a huge range of skills and resources in concerted and organized ways that has made possible the domination of life on our planet that humans, for better or worse, have achieved. This development does not, of course, depend solely on the unique properties of individual humans. The orchestration of effort needed to build a stadium or run a rail network also requires a complex social structure. The evolution of governments, money, educational systems, food distribution networks, and so on are all fascinating stories, but far beyond what I can discuss today. Such institutions shape the processes that result in these massively complex joint ventures. But the developmental flexibility of the human individual is an equally essential background condition. The combination of skills needed to accomplish these massive transformations of our world—niche constructions—is only possible because individual humans have the developmental plasticity to direct their capacities towards a vast range of specialized abilities. Developmental plasticity is characteristic of almost all organisms, but perhaps none to the extent that is found in humans.

Given all this, it is a remarkable fact that political ideology and psychological and economic theory has been dominated for the last 150 years or more by the ideology of individualism. Individuals, in this view of the world, do not generally cooperate, but they compete for a limited quantity of goods. In economics, the most important background condition is scarcity, necessitating this competition. As is well known, the competitive assumptions behind natural selection and economics have been passed backwards and forwards between biological and political thought for over 200 years.[9]

---

[9] See e.g. Gagnier 2000. For a contemporary attempt to continue the interaction, see Hodgson 1996.

Famously, Darwin identified the work of Thomas Malthus, claiming to demonstrate the necessity of poverty or death as the geometric growth of production outpaced the linear growth of resources, as a decisive inspiration for his idea of natural selection.

Not only is the well-being of the individual human generally conceived by central schools of social scientists solely in terms of the satisfaction of specific personal desires, but even the amazing feats of cooperation alluded to a moment ago are seen ultimately as consequences of the choices of self-interested individuals. No doubt the enslaved people who built the White House and the US Capitol in Washington, DC, did so because the option of working on these edifices was preferable to the brutal punishment that was the only alternative. But today, when no one is technically enslaved to build, say, the Burj Khalifa in Dubai, currently the world's tallest building at 828 metres, the cooperation between the army of engineers, construction workers, building supply merchants, electricians, surveyors, painters, plumbers, and so on is generally understood as a concurrence of individually optimal decisions. Even the architects, town planners, and so on who created the conditions that made this cooperation possible are merely pursuing their own personal goals.

There is no doubt a grain of truth to all of this. People in many contemporary countries experience their choices of activity as free, subject, admittedly, to constraints of ability, opportunity, and resource. I can buy whatever I can afford and work in whatever field I have managed to obtain qualifications for (typically subject to resources) and for which there are openings. But this misses a much larger point. Even if I have the resources of Jeff Bezos or Elon Musk, my very existence is dependent on the fact that, for whatever reason, many people are growing food, building houses, etc. As far as I know neither Bezos nor Musk grows food or builds houses, two of the many things that must be provided for me to be free to choose my calling as a philosophy professor or a multi-billionaire. Whatever the significance of the degree of choice we, or some of us, exercise in pursuing our life plans, this autonomy is fairly trivial in comparison to the vast amount of labour of countless others that is necessary merely to maintain our life processes. And of course, this is why it seems so risible for some libertarians to imagine that Bezos or Musk has somehow earned his billions. Even if some good decisions they made have provided great benefits to the human species, the total dependence of the ability to make those decisions on the work of many suggests that a quite modest exceptional reward for their contribution would be sufficient. Even more risible, if possible, is the proposition that only the prospect of wealth many thousands of times that of their less fortunate conspecifics could be sufficient to motivate them to make significant contributions to the general good.

I have pointed to some familiar absurdities in current political discourse, but absurdities that are quite widely believed. Why are such things widely believed? I would like to propose that something that at least makes it much easier to believe is the continued dominance of thing ontology. For a thing ontologist, it is natural to think of individuals as autonomous entities that interact to make a new entity, society; and it is these autonomous constituent individuals that determine what this new entity does. Autonomy is, after all, one of the fundamental characteristics of a thing. Or perhaps, as Margaret Thatcher, a notorious champion of individualist ideology, famously remarked, society does not exist at all. The profound falsity of this remark[10] perhaps helps to explain the incalculable damage that the ideology it expresses has done to the lives of billions. At any rate, some such picture is thoroughly articulated in two varyingly influential scientific programmes based entirely on the interactions of purely self-interested competitive individuals, namely evolutionary psychology and neoclassical economics. I have discussed the intrinsic defects of these projects in detail elsewhere (Dupré 2001), and won't try to do so again here. For present purposes it will be sufficient to stress how much they share the grounding in isolated autonomous individuals that I have been and shall be criticizing throughout this book. Seeing the human individual as a complex process sustained in its far from equilibrium condition by countless interactions with other organisms, human and otherwise, and by a vast infrastructure also created by vast numbers of humans dead and alive, the assumption of isolated self-interested individuals seems nonsensical. Indeed, it is no surprise to find it to be an assumption that serves mainly to foster socially dysfunctional behaviour.

Let me just add that it is not my intent to deny any status to the human individual or reduce human life to the condition of interchangeable ants, devoted only to the service of the community. I shall take up this topic in later chapters, and especially in the final chapter, which addresses individual freedom. In chapter 5 I shall take up the theme I have just been discussing, of human sociality, and explore some perhaps surprising implications for the understanding of various classifications within the human species.

---

[10] At least as naturally understood. The meaning of the famous quote has been much debated. As my own contribution to the discussion, perhaps Margaret Thatcher was a process ontologist: there is no such *thing* as society.

4
Personal Identity

Introduction

Whatever the problems in saying what the boundaries of the organism might be at a time, or the boundaries in time of a stage of an organism, we cannot entirely avoid a question about the persistence of the human organism through time. This is part, at least, of the classic philosophical problem of personal identity. What makes me the same person as a child who existed several decades ago? My starting point for addressing this question is the position that has recently been defended as animalism: the continuity of a person over time is just the continuity of a certain kind of animal. The position is, however, substantially modified when combined with the view that an animal, human or otherwise, is a process. In this chapter, I shall elaborate this synthesis, which I describe as 'processual animalism' (Dupré 2014; Meincke 2018, 2021). I shall further consider what implications this position might have for ideas of personal responsibility and for the possibility of, if not immortality, massive extension of the human life span.

What is it for a person to persist over time? This is an ancient philosophical question. The traditional puzzle can be introduced by way of John Locke's tale of the Prince and the Cobbler, a thought experiment that can be found in various forms in literature and more recently film. Locke imagines a prince waking up in his palace claiming to have all the memories, beliefs, and consciousness of a cobbler; a cobbler wakes up at the same moment in his modest hovel with the memories and consciousness of the prince. The cobbler believes, and declares, that he is really the prince. He feels as if he has just woken up in an alien body in a highly unfamiliar environment. Similarly for the prince, though he may be less insistent on his true identity.

Note that there is no uncontentious way of telling the story. I have told it in what might be described as a third-person way, as it would appear to the prince's valet when he wakes the prince in the morning to discover that he has gone mad, or to the family of the cobbler, for whom the nonsense will also be fraught with danger. But there is a perhaps more standard first-person

Everyone Flows. John Dupre, Oxford University Press. © John Dupré 2025.
DOI: 10.1093/9780198941866.003.0004

perspective. From this point of view, we imagine the prince waking up and finding he is in a wretched hovel and his appearance has changed dramatically. This second description is the one that people generally find more intuitively plausible and indeed this is the one urged on us by Locke. We can imagine, or think we can, waking up in a different body. Intuitively, we are mostly Cartesian dualists: we think of our minds as independent from, and in principle even detachable from, the bodies in which they reside. But I originally introduced the story in a quite different way: two people have gone mad. Only the co-occurrence of the two transformations makes this story less plausible. If a cobbler wakes up and declares that he is a prince who has migrated into this body from somewhere or other, we think he has gone mad. Indeed, thinking one is Jesus Christ, or Napoleon, is a familiar if extreme form of psychosis.

As traditionally interpreted, this problem can be seen as the question, One thing or two? Locke's view, drawing on the common intuitions about the prince and the cobbler, is that identity is constituted by consciousness and memory. This inheres in something distinct from the physical body, a mind or soul. Materialists reject such a dualism and insist that mental properties such as memory inhere in the physical body. While my sympathies are more with the latter view, I take a different approach.

The story does, anyhow, helpfully distinguish the two main traditional approaches to the question of personal identity, the question what it is for a person to persist over time. On one very natural view, exemplified by Locke, memory of one's past experiences provides the feature of the present individual that links him or her to the person who had those experiences. The alternative view, recently increasing in popularity, is that it is the human body that determines identity, regardless of any psychological discontinuities. I just noted the idea that the question might come down to whether there is one thing or two in a human person. In my view there is neither one thing nor two, but none. The human is a process not a thing. So, much of this chapter will concern questions about how we identify and individuate processes. I hope that this will cast somewhat novel light on the ancient debate.

Individuating Processes

I begin with two very general points, one negative, one positive. The negative point is that there is no a priori limit to how much change there can be in the lifetime of a process. I said in chapter 2 that there was a lineage, a process, leading from LUCA, the last universal common ancestor, to the contemporary human species. This was not a claim about any resemblance we might have to an ancient and primitive cell. Positively, the reason why this may nonetheless count as a process is that each stage in it is causally connected to the next. I don't assume, in saying this, that there is any objectively given decomposition of a process into particular stages. Rather, it is just that however we draw a distinction between stages there are causal links that lead from the earlier to the later stages. But how we demarcate an individual within such an extended process is substantially undetermined. This already suggests a view that I shall defend in this chapter: the problem of personal identity is overstated. The phenomenon it aims to understand, the existence of a unique individual over a specific length of time, is much less clear than is normally supposed.

In elaborating this view, I shall start with an increasingly popular materialistic theory of personal identity, animalism. I shall then reinterpret this view in light of my processual commitments. A poster for a conference on animalism a few years ago described the position thus: 'According to "animalism", our fundamental nature is given not by our psychological capacities, but by our biological constitution: we are primates (*Homo sapiens*), and like all organisms, we persist just in case we continue living.'[1] I am generally sympathetic to animalism as thus described. I agree that we are members of the species *Homo sapiens*, and that we persist as long as we continue to live. My reservations concern only the 'fundamental nature'.

Fundamental nature here relates to a widely held philosophical picture of persistence. For a thing to be a possible subject of persistence, on this view, it must be a thing of a certain kind. It persists so long as it continues to satisfy the conditions for being a thing of that kind, but no longer. The conditions are the essence—the fundamental nature—of the kind. Thus, *pace* Kafka, it is impossible to turn into a beetle.[2] Although this is certainly a theory devised to understand the persistence of things, I will not hold that against it. The problem, for biological entities such as organisms, is that whether they are things or processes, there is no such essence or fundamental nature.

---

[1]   'The Lives of Human Animals', Spindel Conference, Memphis, 2013.
[2]   Or any other insect, for those who disagree with the translation of Nabokov's *Ungeziefer*.

As noted in chapter 2, any two species can be connected by a series of species comprising the lineage of each traced back to a common ancestor. Very many properties will change throughout this sometimes very long trajectory. Why should we count one, or a specific set, of these as the essence, the sine qua non, of a species existing at any particular time? There is a serious dilemma confronted by such a choice. An essence should be a property possessed by all and only the members of a species. All is easiest enough to satisfy. All humans have hearts, brains, livers, etc. But so do all vertebrates, probably. Only humans have flown to the moon or written a novel; but most humans have done neither. A property of all and only humans is much harder to find. The variation known to exist within the species, and incidentally a condition of evolution by natural selection, makes it very likely that there is no such property.

Suppose, as is widely accepted, we think the human species is defined by the reproductive relations between its members, actual and possible. Some philosophers have recently proposed that we allow relational essences or, more specifically to the present question, historical essences, a series of reproductive links to an ancestral population. The problem with this is that it is a strictly relational property, and if anything follows about the intrinsic properties of the related members of the species it does so only contingently. But an essence is generally supposed to provide a property necessarily true of any member of a kind; and in the present context, it is a property of the individual that is supposed to determine the persistence of an individual. When a human dies or, for that matter, is transmuted into a porcupine, they do not persist as a human; but they do not then cease to have been born of human parents. So no guidance is provided for our question about persistence.

The unlimited change within a lineage makes it difficult to define an essential property for an individual qua member of a lineage, but the nature of individual biological development presents comparably difficult problems. Consider the life cycle of a beetle, through egg, larva, pupa, and adult. What property do these stages have in common that constitutes them as parts of the same life cycle? One might imagine that the genome provided an answer. But genes mutate, recombine, and change in various ways all the time. Any attempt to find an essential feature of the genome is going to face the same problem in applying both to all and to only members of the species. The sorts of changes that would prevent the continued life cycle of the organism are the kinds of genetic features that will be strongly stabilized by selection and thus likely to occur in many related organisms. Features unique to the species are unlikely to be universal within the species.

The Persistence of Processes

It is time to move away from conceptions of identity honed specifically for a world of things and consider more directly what it is for a process to remain the same process over time. In one sense, persistence is not a problem for processes: that is just what they do. Recalling the definition of animalism above, if an organism does enough to stay alive, it persists. There is no particular requirement of similarity between stages of a process, hence no question for appeal to an essence to answer.

A second point that has already appeared several times in this book, is that it should not be assumed that there is a unique way of delineating processes. If we see life as consisting of more or less stabilized patterns maintained in an enormously complex flow of living activity, then there may well be different persisting patterns to distinguish. I have noted the options of including or excluding microbial symbionts from the definition of a multicellular organism; an ant may be considered an independent individual, or a constituent of the much larger hive process. I have in the past referred to this freedom to define individuals in various ways as 'promiscuous individualism' (Dupré 2011).

To elaborate this point, think of the process made famous by Heraclitus, a river. What is the difference between a river and a tributary, and when is the latter part of the former? The combination of the lower Mississippi and the Missouri is considerably longer than what is now referred to as the Mississippi, but the latter carries more water. Either factor might very well have been taken as a defining characteristic, though in fact it seems that the decision reflects historical contingency. From the Treaty of Paris in 1763, the Mississippi served as the boundary between the British and Spanish empires, whereas Meriwether Lewis and William Clark in 1804 were the first Europeans to travel the length of the Missouri. The Mississippi was surely more salient to European settlers at the time that these nomenclatural decisions became established. The philosophical point, at any rate, is just that there is surely no difference of kind between the two candidates for being the river and the two candidates for being tributaries (the Missouri and the Upper Mississippi).

There are many further questions that might be asked in such a context. How big must a stream flowing into a river be to be a tributary? What are the differences between a stream, a creak, a brook, or a river? If a river widens to become a lake is it still also a river? When does the lake begin? I take it that none of these questions is answered by nature, though we may, if we wish, stipulate precise definitions. It is also surely possible, perhaps likely, that hydrologists, cartographers, water engineers, and hikers may make different distinctions. The encompassing process is the drainage of a certain area of land to the sea. The parts of it—rivers, streams, lakes, tributaries, distributaries, and so on—are ways of distinguishing parts of this process to serve our various interests in the process.

We might also ask where the river begins. The Mississippi River is said to begin as a trickle flowing out of Lake Itasca in northern Minnesota. Is there some fact that determines this starting point? Surely not. Why should it not run through Lake Itasca, as it does, for instance, through Lake Winnibigoshish? The importance of this question harks back to my discussion of pregnancy in the last chapter. We should not assume that the beginning of an individual process is a matter of objective fact, either for a river or for an animal. We have reasons for defining these beginnings and different concerns may lead us to name different sections of a process. There is no biological answer to the question when a person comes into being. Conception is where two processes intersect, and the intersection is surely consequential. But just as at the confluence of the Mississippi and the Missouri we may say either begins or ends and the other continues, or two entirely new rivers begin, so with a life. The egg precedes fertilization, and we could perfectly well trace the human life back to the formation of that egg cell. The tiny sperm embedding itself in the relatively enormous egg no more necessitates us thinking of the latter as ceasing to exist than does a tributary running into the Mississippi or, for that matter, a tree falling into it, force us to say that the river has ceased to exist. If we want to address the ethical, social, and political questions about what rights we reasonably attribute to a developing and unborn human and to a mature female human when these conflict, we should not look to biology for an answer.

Similar questions arise with regard to processes of many kinds. Sometimes a difference of degree is sufficient to become a difference in kind. In meteorology we distinguish a tropical disturbance which, as wind speeds increase, becomes a tropical depression, a tropical storm, and finally a hurricane. Indeed, whether it is termed a hurricane, a typhoon, or a cyclone depends only on what body of water it originated in. In biology distinctions tend to be more complex. Consider again the lineage. Any biological lineage can be traced all the way to LUCA, the last universal common ancestor. Along this journey we distinguish it into many kinds, kinds which increasingly deserve to be called species. Are the decisions between these species an objective matter? I think not. If one is a cladist then there is a necessary condition, a branch in the tree, that the origin of a new species must meet. But how big a branch? A small band of animals go and live alone on an island, evolve a little, and die out. Must we rename the originating species? I don't think so. In practice most taxonomists are unlikely to accept the necessary condition without qualification. If a species exists for aeons without branching but undergoes major changes, most will acknowledge the possibility of so-called anagenetic speciation, speciation within a non-branching lineage, as opposed to the cladogenetic speciation championed by strict cladists.

And then think of an organism. The life cycle of an organism is a single process. But we are not obligated to say that at every stage of the life cycle we have the same kind. The egg is part of the same process as the tadpole and the frog but it is arguably a series of different entities: an egg hatches into a tadpole; it is not a tadpole. And there are other distinctions that apply within an organism's life cycle. Large predatory fish can lay millions of eggs; the eggs and the tiny fish that hatch from them are prey from the perspective of their parents, who are as likely to eat them as they are any other organism of similar size. The point where a prey animal grows into a predator is vague. Most organisms are prey for some species, predators for others, and their relations can change over the life cycle. But this is, nonetheless, a distinction without which population ecology would be impossible.

Individuating Human Processes

So let us return to human processes. An immediate thought is that this talk of rivers and fish is all of marginal relevance to the human case. Apart from the question of symbionts, about which one might plausibly not care one way or another for everyday purposes, the uniqueness and continuity of a human person is not contentious. People are born, named, given various identifying numbers, and tracked throughout their lives by various bureaucracies until their deaths are certified. There is no underdetermined decision to be made as to whether I am the same person as the John Dupré who grew up in southern England in the middle of the last century. The question is just what this means. Identity over time is simply a given.

There is undoubtedly a process that leads from fertilized egg through embryo and foetus to baby, child, teenager, and so on to death. Similarly, there is a process that leads from LUCA to the human species. Here, we say there is one process, the lineage, but many different species ordered temporally along the lineage. In the human case, we may say that the embryo, foetus, baby, and so on form one human life cycle. One might think that the move to the person was just what animalism should suggest: an animal is, I hold, a life cycle. But that can't be right, as it would imply that the zygote was a person. This might be acceptable from a religious perspective in which fertilization of the egg is a moment of ensoulment. But for the animalist who believes that the human is only an animal, surely this is too early for the animal to become a person? The concept of a person is generally connected with psychological properties such as memory and consciousness and legal status such as citizenship or bearer of rights and duties. None of these makes much sense in relation to a single cell.

Here we have identified a major qualification of animalism. We cannot identify the human person with the human animal unless we are willing to accept that the zygote and every subsequent stage of the human life cycle is a person. Perhaps we shall have to bite this bullet, but we have already seen that the logic of process does not require it. Generally, and more formally, if some part of a process of kind A is also a process of kind B, it does not follow that every part of the process is of kind B. (This innocuous principle, incidentally, is also true of things, if there are any. My copy of Plato's complete works is currently a door stop; it will no longer be one when the role is taken over by a copy of *Principia Mathematica*. If there are things, I suppose books are things.)

It is, of course, a hugely controversial matter of debate at what point a foetus becomes a person with rights. Ronald Reagan once claimed that: 'Unless and until it can be proven that the unborn child is not a living entity, then its right to life, liberty, and the pursuit of happiness must be protected.'[3] I think we can safely concede that the unborn human is a living entity, though 'child' is a far more controversial description. But we should certainly not accept that every living entity has a right to life, liberty, and the pursuit of happiness. We can debate whether these rights extend to pigs and chickens; but if we grant them to carrots, cauliflowers, and even cholera bacteria, we will be revoking our own right to exist.

So if designating a phase of a process as an X, and thereby legitimating judgements of distinct parts of a process as the same X, is recognized as a pragmatic decision answerable to particular human goals, what are the relevant goals for calling a part of the human life cycle a person. Why, in other words, do we individuate humans at all?

---

[3] <https://www.reaganlibrary.gov/archives/speech/remarks-annual-convention-national-association-evangelicals-orlando-fl/>

Answers to this question are not hard to find. To begin with, we humans individuate one another for organizational purposes, for the distribution of responsibilities and rights essential for the functioning of the complicated structures that support the very large populations of humans. Here we see a major difference from the much simpler division of labour in an ant colony. When a pheromone trail directs foragers to a food source, it seems unlikely that it matters which particular foragers follow the trail; we assume they are pretty much fully interchangeable. Human workers have more specific and differentiated occupations. A human is not merely a retail worker, but one who works at a particular store on the early morning shift. Having retail workers show up at any convenient store at any arbitrary time would probably not be a successful way of managing the goods distribution function. Other aspects of human societies require, for instance, that the same individual who acquires a debt or orders a product is the one who must pay the debt or expect to receive the product. This is managed by uniquely individuating humans and through often quite complex practices of re-identifying them.

There is, of course, a more private, as we say 'personal' or 'personalized', dimension to all this. First of all, individual humans do appear to have a special concern for what happens to themselves, something they may even share with the worker ants, though this can be debated. Interestingly, nonetheless, we do not universally applaud this concern. Selfishness is considered a moral flaw, though a complicated one as we generally acknowledge that to some degree it is not only expected, but perhaps even deserves our approbation. More importantly, perhaps, selfishness is a highly variable feature of humans. In contemporary capitalist societies where selfishness is expected and sometimes even applauded, some individuals seem surprisingly deficient in this character trait. Moreover, it is plausible, though not always easy to assess, that in many other historical, and perhaps even contemporary, societies selfishness was either unusual or perhaps even unimaginable.

Humans generally have strong bonds of affection to specific other humans, especially family members. Perhaps, as many evolutionary theorists argue, the bonds to family are a deep part of our biology. There is no doubt that such bonds are part of our natural history, though there is a good deal of disagreement concerning how this aspect of our biology is sustained (see e.g. Emlen 1995, Davis and Daly 1997). And it is hardly news that family relations are sometimes irreparably broken. Still more controversial in terms of evolutionary origin, but no less indisputable, is the capacity, albeit highly variable, of humans to form a network of bonds beyond the immediate family (Clutton-Brock 2009). At any rate, the network of personal relations that most people value, familial or otherwise, is based on individuality: relations are with specific, not easily replaceable, individuals. Most of us would be unhappy if random strangers regularly showed up for family dinner rather than the familiar loved ones, even if they had exactly the traits we value in our loved ones—though it is not an impossible ideal, that we might come to welcome such strangers as warmly as our relatives.

So, while a part of the importance of our continued identity as persons may be an expression of our biology, notably in the parental care typical of many societies organized into family groups, a large part of this importance is also social. We have complex and pervasive technologies the entire point of which is to track personal identity. We all, or almost all, have names and social security numbers; most of us have telephone numbers, email addresses, and a totally unmanageable number of passwords for accessing our internet spaces. We issue birth certificates, death certificates, identity cards, and passports; many of us now carry quite sophisticated face-recognition and fingerprint identification technologies in our pockets.

If you believe that you have an individual essence, or a unique soul, then these technologies are merely reflections of the contingent impossibility of disclosing our essence or our soul. But if, as I have tried to persuade you, you are a process, constantly changing and with no clear limits to the possibility of change, you should see these technologies as contributing to the construction of identity rather than merely tracking something that pre-exists.

Of course, I don't deny that the human animal has a life cycle that coincides in part with the life history of a person. This is the clear attraction of the animalist account of personal identity. But my question was why we decide to track a person through a very large part of this process, typically from birth when names are given and the first identity-determining paperwork is issued. This seems so natural as to be inevitable, but surely it isn't. Imagine a society of intelligent frogs. Female frogs, let's imagine, deposit a predetermined number of eggs in the communal pool, teachers and carers look after the tadpoles, and when these turn into frogs they are given names and social roles. A frog, one imagines would have no great interest in the tadpole, still less egg, from which it emerged.

But we do not need to fantasize about intelligent frogs. Human life is already more variable than those of us who have internalized and naturalized Euro-American culture assume. Consider the distinction between what the anthropologist of childhood David Lancy calls 'pick when ripe' cultures and 'pick when green' cultures:

> In the '**pick when ripe**' culture, babies and toddlers are largely ignored by adults, and may not be named until they're weaned. They undergo ... a 'village curriculum': running errands, delivering messages and doing small-scale versions of adult tasks. Only later are they 'picked,' or fully recognized as individuals. In contrast, in '**pick when green**' cultures, including our own, it's never too early to socialize babies or recognize their personhood. (Lancy 2014)

Lancy proposes that the 'pick when green' culture emerged in the Netherlands in the seventeenth century as a reflection of radical Protestantism, and was imported to the English speaking world by John Locke. These children are the 'cherubs' in the title of his book, *The Anthropology of Childhood: Cherubs, Chattel, Changelings*. Despite the discomfort they may cause modern Western readers, the other two C-words better reflect historical and cross-cultural norms.

In relation to my present concerns, the point is more modest. It is neither universal nor even culturally typical that personhood begins at birth—or, for that matter, as in some versions of Christianity, long before birth. Again, the point is reinforced that personhood is a concept imposed on the animal life cycle for variable parts of the life cycle and reflecting wider social interests.

These reflections may well be seen as something of a debunking of personal identity, and I think that is correct. There is a process, part of an animal life cycle, that led to whatever I am today. What relation should I feel to past stages of that process? Perhaps I should feel pride in the achievements of a past self and shame for my misdeeds. But is this very different from the pride or shame I feel in the achievements of my friends and family? I like to be associated with good things and not with bad things.

Here I converge on conclusions reached, for very different reasons, by the late Derek Parfit. Parfit doesn't think that personal identity is important. Parfit thinks, like Locke, that what matters is the connection between one's present experience and one's memories. But he doesn't think that this is enough to ground an interesting relation of personal identity. There are parts of my life about which I remember very little. Unless I decide to write an autobiography, the person who experienced this life stage will be of no special interest to me.

Parfit finds the unimportance of personal identity liberating. He writes that when he believed that personal identity was what mattered, he was more concerned by the inevitability of his death:

After my death, there will be no one living who will be me. I can now redescribe this fact. Though there will later be many experiences, none of these experiences will be connected to my present experiences by chains of such direct connections as those involved in experience-memory, or in the carrying out of an earlier intention. Some of these future experiences may be related to my present experiences in less direct ways. There will later be some memories about my life. And there may later be thoughts that are influenced by mine, or things done as the result of my advice. My death will break the more direct relations between my present experiences and future experiences, but it will not break various other relations. This is all there is to the fact that there will be no one living who will be me. Now that I have seen this, my death seems to me less bad. (Parfit 1984, 281)

Certainly, my past can have consequences for my present self. If an earlier stage of my animal life cycle had written a best-selling book, perhaps I would still be receiving royalties from it. And if it had committed heinous crimes, I might still be held responsible and punished. The latter case is an interesting one. There is a strong and very understandable sentiment in favour of holding octogenarian former war criminals accountable after many decades. But there is surely also an element of discomfort in such actions. The octogenarian may have spent his life doing good works in an effort to repay his moral debt. The worry is not that perhaps he has repaid the debt; the lives of hundreds or thousands of victims cannot be restored. It is that the person now living may now actually be a morally admirable person. This could be debated. Perhaps if he had achieved moral excellence, he would long ago have turned himself in for his crimes. But whatever the upshot of this discussion, there is surely no doubt that this is a very different situation from arresting someone red-handed in the commission of their crimes. The relation to once distant precursors is more complex than mere identity.

Toxic Heritocracy

Let me mention one more reason why I think it would be very beneficial to diminish the importance we attach to personal identity over long periods of time. One of the most serious problems of our time is the obscene and rapidly increasing inequality between people in most countries of the world. Some of this derives from the ways, to which I have already alluded, that people are considered entitled to what are judged to be the fruits of their labour. But the ratchet that keeps inequality increasing is the chains of entitlement that run through human lives and then are passed on to offspring. Many of the grossly rich are so because an ancestor several years back became rich, perhaps in his or her youth. The vehicle that carries from the events that made this ancestor rich to the current plutocrats is the absolute identity of persons carrying absolute rights to property. If one relaxed the absoluteness of identity, one would naturally also relax the absoluteness of these rights, and it would be difficult to justify the extension of these not merely to the end of life but beyond through rights of inheritance.

The heritocracy, as this dysfunctional grounding of inequality is sometimes known, is an extreme expression of the individualism that has been an occasional target throughout this book. The quote above from Parfit also points to a weakening of the primacy of the individual in favour of a clearer recognition of the profoundly social nature of our species. In addition to the dependence of the individual on countless other humans, every human is a member of a range of more or less significant social wholes: families, villages, tribes, armies, nations, universities, bridge clubs, gangs, and so on. Often the individual is better understood as a part of such a whole than as an individual process. Recalling the hierarchy of processes that constitute the living world, and the ways in which processes are stabilized both by their parts and by the wholes of which they are parts, we should not let the concern with our own status as individuatable processes lead us to overlook other processes of often even greater significance.

(Near) Immortality

Picking up on Parfit's remarks on mortality and the consolation of philosophy, let me end with a brief reflection on a topic that has attained some prominence recently, the possibility of, if not immortality, at least a massive extension of the duration of life. The traditional conception of human immortality supposes the human essence to reside in the soul, a substance wholly different from the human body that we know to be so sadly perishable. I must admit that I do not fully understand this doctrine. I must also confess that if it is true, then most of what I have said in this chapter is false. The soul is, I suppose, an immutable substance, and it has an essence, a category of which I have been highly sceptical. I will not, anyhow, consider this possibility any further here.

Perhaps an unusual way into the topic of immortality is to recall the chapter of Richard Dawkins's famous book, *The Selfish Gene* (1976), entitled 'Immortal Coils', which argues for the immortality of genes. Tempting though it is to have some fun with the appeal to immortality by such a notorious atheist, fairness to Dawkins requires me to note that the immortality in question is the immortality (or considerable longevity as we should surely more accurately describe it) of gene lineages, processes in which the material genes are constantly replaced, rather than gene individuals.[4] And this provides a valuable reminder that the persistence of processes differs in many significant respects from the persistence of things.

Somewhat parallel to Dawkins's account of the life of a gene, there is a relatively uncontroversial sense in which one might achieve immortality through one's offspring. This is not so different from the immortality that an Aristotle, Shakespeare, or Einstein may attain through their works: the world in the distant future may contain an identifiable causal trace of one's existence. This is also part of what Parfit refers to in his more sanguine view of mortality. But although it is something some people may care deeply about, it is hardly to be confused with the idea that the very process which constitutes one's life might exist for a long period after one's death. The difference has been succinctly stated by Woody Allen: 'I don't want to achieve immortality through my work; I want to achieve immortality through not dying. I don't want to live on in the hearts of my countrymen; I want to live on in my apartment.'[5]

More exciting for those who find themselves dissatisfied with the brevity of the human lifespan are the pronouncements of some who believe that medical science will soon be able to overcome the diseases of old age and to extend our lifespan by an order of magnitude (De Grey and Rae 2007). This seems to be an especially pressing concern for some of the multi-billionaires I referred to earlier, some of whom fund this research; perhaps they feel it to be unjust that their riches are not yet sufficient to buy immortality. If one thinks of the human organism as an object or machine subject to deterioration, this is at least a reasonable aspiration. It is characteristic of the machine view of the organism that one thinks of a period of development, in which

---

[4] One might be tempted rather to contrast the immortality of gene tokens and gene types. However, a type, I take it, is an abstract entity, and if it has any temporality it would have to be immortality. Better would be the unceasing instantiation of the type; but at this point it is surely better to focus on the process by which the type continues to be instantiated.

[5] Quoted in Lax 1975. Thanks to Gabe Dupre for alerting me to this delightful quote.

the machine is constructed, a stable period of mature function, and a period of decay. The objective is to maintain the machine in the middle phase by counteracting the various causes of decay or malfunction. Protection from sources of damage and repair or replacement of seriously damaged components should, in principle, enable a machine to be maintained indefinitely. However, recognizing the human organism as a kind of process makes this story considerably more problematic.

A first worry concerns the possibly confused motivation for seeking many centuries of life. Processes do not persist by sustaining an unchanging and potentially immortal essence, but rather by the continuity and interconnectedness of the activities that maintain them. So does it make sense for me to be concerned about the existence in a thousand years of some person whose connection to me is merely historical, with whom I have little in common beyond mere shared participation in a continuous process? I have questioned the significance of identity between a living adult and some dimly recalled child many decades earlier; one may equally doubt the possibility for a child really to imagine that they will one day be an old person, and the consequent dubiousness of supposing the child to have any real concern with the coming to be of any such antiquity. The concern of myself today with the future existence of someone processually continuous with me in the distant future, but who will long ago have forgotten the thoughts, concerns, and memories that now absorb me, is an extreme scenario for similar doubts. It is true that I can imagine that there will be someone who shares my social security number and has the same name as me on their passport. But why should I care about this?

This relates to a further problem, which I mentioned briefly in chapter 1. As I have insisted, the human, as an animal, is fundamentally a life cycle. Stabilizing one temporal stage of the life cycle (one hopes a relatively youthful one!) for a period many times the current lifespan of the entire cycle would be to create a fundamentally different kind of being. As the procedures envisaged for life extension are often difficult and dangerous it is perhaps likely that they will not be undertaken until natural life is almost exhausted, leading to the familiar objection that the outcome of these efforts will be a dystopian world populated entirely by the aged. In the worst case, the ageing process continues modified only by the absence of death, so the degree of antiquity reached is something beyond our current imagination.

Some transhumanists who promote the search for immortality have an idea that at least evades these worries. They embrace the rather repulsive parody of Cartesian dualism in which we continue our eternal lives as programs in computers, certainly a more radical ontological shift away

(a)   (b)

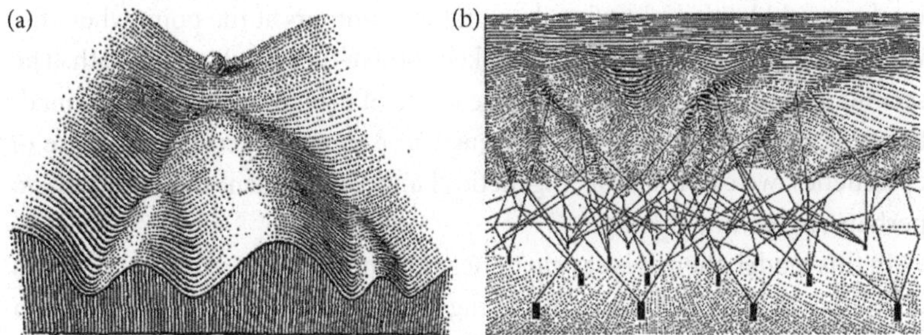

Figure 4.1 An epigenetic landscape. From Waddington (1957). Drawing by John Piper (1903-1992).

from a living process.[6] The sense in which I, a material animal, would be continuing my existence as code or programs running in silicon chips does, I must confess, entirely elude me. It is perhaps no more impossible than that I might continue my existence as some kind of stream of consciousness in an immaterial substance. But since it is no less impossible, I feel comfortable ignoring it. Even if such a thing could exist,[7] the ontological disjuncture between what we are and what it is supposed we might become strongly suggests the impossibility of continuity of process between the two.

Let us return to the ontologically less bizarre proposals for material persistence for millennia. As I have stressed from the outset, for a persistent process such as a human, a question that can always be raised is what maintains its more or less stable structure. As I mentioned in chapter 1, the stability of an organism is, in terms used by Waddington (1957), homeorhetic rather than homeostatic. A specific trajectory rather than a steady state is what is maintained. This is the kind of process illustrated by Waddington's famous diagrams showing a ball rolling down a valley (figure 4.1 (a)), which he described as epigenetic landscapes.

The slopes on either side of the valley represent the processes that keep the ball in the valley or, as interpreted, keep a life cycle on a normal trajectory. Dysfunctions, for example cancer, could be represented by one of the branching valleys. And healthy or unhealthy modes of behaviour or dietary

[6] Sandberg and Bostrom 2008. Following Bostrom's (2003) notorious simulation argument, there has been considerable discussion of the idea that we probably are now simulations in computers.

[7] See Mandelbaum 2022 for why it couldn't.

intakes could be thought of as changing the contours at the point where the valleys meet, in ways that affect the likelihood of the ball changing path. The second diagram (4.1b) shows how the shape of the landscape is maintained; the pegs and guy ropes represent genes. In a more contemporary sense of epigenetics, we can think of epigenetic changes to the genome as cutting or, perhaps, tightening the ropes.

In this model, then, it is easy to understand how appropriate interventions could maximize the chances of staying on the healthy developmental path for the full lifespan. That is what medicine attempts to do. But a ball cannot roll downhill for ever; and likewise it is not obvious that the developmental trajectory of a human can be extended forever. I do not want to argue here that this is demonstrably impossible. However, the misguided analogy with maintenance of a machine makes the plausibility of indefinite life extension much less problematic than would a more accurate picture of a process with a characteristic trajectory and a typical temporal duration.

Let me conclude. As I have stressed throughout this and earlier chapters, processes do not divide the world in the precise countable way expected of things. And though we do not generally see the individuation of humans into distinct persons as a deep practical problem, questions about when person-hood begins or what it would be to extend it indefinitely in the future do raise indeterminacies that are not present in the world of substance. But perhaps the real enlightenment provided by process ontology is the recognition of the interweaving and mutual stabilization of processes of which the human animal provides an outstanding illustration. A human is in part a cell lineage, a functional living system, a psychological subject, and a part of many social wholes. Which of these must persist if the human person is to persist? I am sceptical whether we can give answers to this question that are more than pragmatic solutions to particular problems.

5

Human Nature and Human Kinds

Human Nature

In this chapter I turn, first, to the age-old question of human nature. Is there even such a thing? In the light of my processual view of life, I shall reject all but the most minimal conceptions of human nature. I then consider the question of divisions within the human species—sex, race, nation, and more, and ask how we should think about these. Again, a process perspective provides distinctive insight into the questions posed by these controversial categories. Both questions have sometimes been addressed from a physiological perspective. My main concern here, however, will be with understanding human behaviour.[1]

The human species, as I have noted in a previous chapter, is an exceptionally well-delineated lineage. We have no close relatives—probably because we wiped them all out—and so there is (almost certainly) no hybridization between humans and any other species. The set of organisms (or perhaps, strictly, holobionts) that are members of the species is entirely determinate. A necessary and sufficient condition for being human is to be part of this lineage, or at least a recent segment of it. This has led some philosophers to a rather simple account of the human essence: it is a historical essence, the property of being descended from a particular ancestral population.

I have no deep objection to such a historical essence, except for some discomfort with the word 'essence'. Essences serve a range of philosophical functions. One is to distinguish what is and isn't a member of a kind, and the historical essence serves this task admirably. But a second function is to tell us something about the members of a kind, something reliably true of all of them. And here the historical essence fails miserably. If, through some bizarre series of mutations, a woman gave birth to a rabbit or an armadillo,

---

[1] Given the enormous amount that has been written about human nature, from extremely diverse perspectives ranging from psychology and anthropology to philosophy and literary criticism, I shall not attempt to relate the following brief discussion to much of this background. For recent discussion from my own field, the philosophy of biology, see Hannon and Lewens 2018.

Everyone Flows. John Dupre, Oxford University Press. © John Dupré 2025.
DOI: 10.1093/9780198941866.003.0005

that would, in the light of a historical essence, count as human. In principle, the historical essence places no limits at all on the properties that an entity that shares it might have. Such an essence tells us nothing about human nature, the goal that seekers after the human essence are generally concerned with; they have especially been concerned to find the key to a deeper understanding of the ways humans behave. Here I shall consider only more robust approaches to human nature that might help to provide such understanding.

Recent philosophy and science have seen heated debates on this topic. Biological theories of human nature have, unsurprisingly, followed the dominant biological theories of our time. Foremost among these is evolution, and evolutionary theories of human nature, first sociobiology, and then, with a slight rebrand, evolutionary psychology, have been much promoted and debated. Perhaps equally prestigious in more recent biology has been genetics, and sure enough there has been considerable effort devoted to seeking genetic causes of behaviour. Since dominant theories of evolution have stressed changes to genes as underpinning evolutionary change, these approaches have generally been taken to be readily compatible.

Social scientists and philosophers, however, have generally been hostile to both evolutionary and genetic explanations of behaviour.[2] At the root of this hostility is the view that humans are far more subtly and complexly responsive to their environments than such explanations can allow. Defenders of the biological perspective, most notably Steven Pinker, have responded by accusing their opponents of treating the human mind as a blank slate, on which society can write anything it chooses (Pinker 2003).

There is, however, a curious kind of semi-truce in this debate. Both sides agree that what we need is a so-called interactionism, which recognizes human behaviour as resulting from a mixture of biological and environmental, especially social or cultural, factors. But the truce is somewhat illusory. For both sides tend to claim that interactionism is exactly what they have always proposed. It is their benighted opponents who insist on their biological determinism or their blank slates. The problem, then, seems to be how to articulate a plausible interactionism.

The most prominent biological theory of human nature at present is Evolutionary Psychology, especially as propounded by anthropologist and psychologist John Tooby and Leda Cosmides (Tooby and Cosmides 2005; Barkow, Cosmides, and Tooby 1995), sometimes known as the Santa

---

[2] Detailed philosophical critiques of Evolutionary Psychology include Dupré 2001 and Buller 2006.

Barbara School after the university at which both worked. This is not, of course, merely the banal thesis that we evolved and so, therefore, did our psychology; it is a specific account of how our evolution determined our psychology.[3] The basic thesis of Evolutionary Psychology is that we evolved a set of mental modules to adapt to life in the Stone Age and these ultimately drive our current behaviour. Recurrent problems faced our distant ancestors concerning such things as access to food or sexual partners or negotiations with other humans, and psychological mechanisms evolved to generate optimal responses to these problems. The Stone Age, or more precisely the Pleistocene, covers the last 2.5 million years, excluding the last 12,000, generally referred to as the Holocene. Its relevance is that it covers a long period of distinctively human evolution after our split from our cousins the apes. It is also, more controversially, suggested that the human environment during this period was relatively stable, allowing for sustained selection of optimal traits. Modern human behaviour has only existed for a very short time and not long enough, Evolutionary Psychologists claim, for significant evolutionary change to have occurred.

Numerous objections have been raised to the Evolutionary Psychology picture, many focused on underlying disagreements about the nature of the evolutionary process. The atavism of Evolutionary Psychology, the idea that the roots of human behaviour were shaped in a distant past age, depends on the idea that evolution is wholly genetic; and the accumulation of the genes supposedly sufficient to shape behaviour would take thousands of generations. But more expansive views of evolution include other forms of inheritance, such as cultural or epigenetic, and taking account also of the changes to the environment produced by the evolving population, niche construction, these can promote far more rapid changes.[4] If these processes are recognized, the underlying causes of human behaviour can be seen to have a much more recent origin, to be more specifically adapted to the present, and, crucially, may vary in major ways from time to time and place to place. It is the actual observation of such historical and cross-cultural variation by historians and social scientists that has done a great deal to drive these arguments.

In endorsing these objections, which I do, am I endorsing a blank slate view? Not at all. Very few thinkers, I suggest, have ever really believed any

---

[3] It has become common to refer to the Santa Barbara school as Evolutionary Psychology, with capital letters. Other attempts to apply evolution to aspects of human psychology and behaviour are then referred to just as evolutionary psychology, lower case.

[4] This more extensive view of inheritance has been especially championed by the Israeli theoretical biologist, Eva Jablonka (Jablonka and Lamb 2014).

such thing. To mention a few objections: humans strongly resist some forms of enculturation, for example enslavement; no one seriously doubts that humans are highly social, typically enjoy status, food, sex, etc.; and there are surely psychological constraints on what kind of social organization is even possible.

Given that Evolutionary Psychologists generally propose that what evolved were psychological mechanisms or modules that drive behaviour only in response to the observed environmental conditions, is there perhaps a convergence on a viable interactionism? Perhaps some, but not all that much, I fear. The difference, I think, is that while not quite a blank slate, the critics of Evolutionary Psychology do generally believe that a distinctive feature of humans is the flexibility and adaptability of their behaviour—not unlimited flexibility, but a lot more than Evolutionary Psychologists generally allow. In chapter 3 I mentioned how this made possible the unique division of labour in human societies, that different members of a society develop very different suites of skills. These are surely grounded in a physical structure, principally the brain, but probably not in modules dedicated to specific aspects of Pleistocene life in the ways Evolutionary Psychology has assumed.

One reason to believe in this flexibility is that, as I mentioned, it is probably quite false that the Pleistocene was a constant environment. What made humans so successful was not their close fit to these putatively stable conditions, but their adaptability to different conditions, perhaps especially those that came about at the beginning of the Holocene and end of the Pleistocene, the end of the last Ice Age 12,000 years ago. This growing adaptability is a far more plausible explanation of the extraordinary successes of the human species than is the evolution of a set of more or less stereotyped responses to the supposedly fixed conditions of Stone Age life.

Human Plasticity

Strongly concordant with the preceding point is my continuing argument that we should see humans as open-ended developmental processes. Plasticity, I have previously observed, is a common feature of the great majority at least of multicellular organisms.[5] For an organism for which complex

---

[5] The *locus classicus* for the general topic of developmental plasticity is West-Eberhard 2003.

behaviour is as central to its way of life as it is for humans, it would be re-markable if behaviour were not a flexible outcome of the very long develop-mental process that humans exhibit. Contra the blank slate, plasticity is not unlimited. Contra the Evolutionary Psychologists, we have very little idea of the limits of human developmental plasticity. Pragmatically, given this situ-ation, we should remain agnostic about the limits of human possibility and explore ways of extending them. This does not require a commitment to the blank slate.

Human plasticity is especially important in moderating our tendency to ascribe differences in human development to differences in innate ability. This point about human plasticity has been beautifully described by a his-torical figure who will be surprising to some in this context, Adam Smith. Smith wrote:

The difference of natural talents in different men, is, in reality, much less than we are aware of; and the very different genius which appears to distin-guish men of different professions, when grown up to maturity, is not upon many occasions so much the cause, as the effect of the division of labour. The difference between the most dissimilar characters, between a philoso-pher and a common street porter, for example, seems to arise not so much from nature, as from habit, custom, and education. When they came into the world, and for the first six or eight years of their existence, they were, perhaps, very much alike, and neither their parents nor play-fellows could perceive any remarkable difference. About that age, or soon after, they come to be employed in very different occupations. The difference of tal-ents comes then to be taken notice of, and widens by degrees, till at last the vanity of the philosopher is willing to acknowledge scarce any resemblance. (Smith 1982/1776, book 1, chapter 2.4)

I think this is just right. The differences between humans, and between dif-ferent categories of humans, are not primarily to be looked for in the playing out of genetic fortune or misfortune, but in the vastly different circumstances in which different humans develop.

To summarize the conclusions about human nature, the reason this is a generally unhelpful concept is that both the human lineage and the human individuals that comprise it are highly plastic, adaptable processes. We can, of course, make statistical empirical investigations about what humans are like at this time and in a particular place, and there are often good reasons to do so. The wider we spread our net over human populations, however, the more variance we are likely to find in our investigations. And the further we project these generalizations into the future, the less reliable they are likely to be.

For the remainder of this chapter I shall explore human variability by looking at some of the finer structure of the process that is the human species. Humans are extremely diverse in their habits, behaviour, and life styles. Being humans ourselves, this diversity tends to be of considerable interest to us, and leads us to multiple ways of dividing ourselves into kinds—age, occupation, social class, race, gender, and so on. However, we should generally be suspicious of strong claims about the implications of belonging to such human kinds. One pervasive reason for this is that the features that form the basis of these classifications generally have multiple causes, and members of the classes distinguished may therefore differ in multiple other respects. A more subtle point that has been persuasively developed by the philosopher Ian Hacking is that much of the homogeneity within human kinds is actually a consequence rather than a cause of the classification. In these cases, which Hacking referred to as 'looping kinds', characteristic behaviour is largely a consequence of the distinguishing of the kind (Hacking 1995). A classic example of this phenomenon is the emergence of homosexuals as a kind of person, rather than homosexuality as a kind of behaviour more or less favoured by some people, as has been proposed and documented by Michel Foucault (1997). Gay culture, gay affect, and other characteristics that enable some to immediately identify some gay people are not natural concomitants of being gay, pleiotropic consequences of imagined gay genes, but a suite of behaviours that emerged as people responded to being classified as a distinct kind.

Sex and Gender

Let me start with the distinction, or family of distinctions, among people that has the strongest ties to biology, that between the sexes or genders. It may help to begin with some elementary biology. Biological sex is understood by biologists as a mating system which, as I discussed in an earlier lecture, is also evolutionarily significant for its role in creating integrated, stable lineages. It consists of a division into two mating types, each providing one gamete (e.g. sperm or egg), the two together forming the zygote that launches the life cycle of a new organism. The gametes are usually of very different sizes, and the sex that provides the larger gamete is usually called female, that which provides the smaller gamete is called male. In the human case, the female, in addition to providing the larger gamete, the egg, carries the physiological adaptations for gestation, birth, and lactation. The genetic difference that correlates with physiological sex differences in humans is the possession of XX (female) or XY (male) chromosomes in addition to the twenty-two pairs of so-called autosomal chromosomes. This genetic difference does not apply generally. Birds have a quite different genetic differentiation. Many reptiles develop into one sex or the other depending on ambient temperature. And some fish change sex as they reach a certain size. The phenomenon of sex difference, gamete size differentiation, must not be confused with any of its various correlates. And finally, though I shan't go into this here, there are duplications and deletions in the sex chromosomes, so not everyone is XX or XY, and not everyone with XX or XY chromosomes follows the sex-typical developmental path. Nothing in biology is simple.

So far the basic biology. This is not very far at all in discussing the natures of the kinds that sexual difference defines for humans. This is fairly obvious based simply on observation of the huge historical and cross-cultural differences in behavioural norms for the sexes. Just compare the anthropological record on polygyny, polyandry, and monogamy, only the most basic distinctions among marriage kinds. Much better to start with something like sex/gender, as the feminist biologist Anne Fausto-Sterling has named it. Sex/gender is a roughly bifurcating developmental process, that leads from the sex assignment at birth, usually but not invariably aligning with the XX or XY genotype, to gendered adults. The gender-differentiated characteristics of mature adults in most or all societies are highly normative, and many aspects of the developmental infrastructure are explicitly directed to achieving this outcome. Many factors—genes, hormones, brains, parenting, schools, social networks, etc.—contribute to stabilizing this process. Nonetheless, there is variation all along the trajectory and there is a diverse set of possible outcomes. A central aspect of this diversity is acknowledged and given some degree of normative resistance to the dominant narrative, with the acronym LGTBQ+ or its variants.

How should we understand the diverse outcomes of this gendering process? It will be obvious that I can only scratch the surface of this question here. The central point relative to my current project is to stress that sex/gender is a developmental *process*. Substance thinking divides the world into kinds of things with characteristic properties. A zygote is a thing of one of two kinds, male and female, and it has a set of mechanisms that normally guide the unfolding of a certain kind of adult (through the genetic *programme*).

Substantially different outcomes are generally considered pathological, as has indeed often been the case for the medical history of diversity in gender or sexuality. (Alternatively, putting even more weight on the normativity, they are considered moral failures.)

Until quite recently, it was standard to think of sex as the fixed starting point and gender as the culturally variable and contestable developmental outcome. This is a framework that has led to a great deal of insight from the work of feminist scholars over the last fifty years and more. In particular, it has increasingly undermined the picture of gender as a natural unfolding, as it has both revealed the empirical diversity of normative gender systems and disclosed the processes by which particular normative systems are implemented. A striking example, though one among thousands, is provided by a series of photographs taken by the Korean artist JeongMee Yoon of children with their clothes, toys, and other belongings.[6] Those of boys are almost uniformly blue, while the possessions of the girls are homogeneously pink. The effect is striking.

One might, I suppose, imagine that it was a natural feature of people with an XX chromosome that they have a liking for the colour pink. This seems unlikely, given that pink has been a colour associated with masculinity until quite recently. Yoon quotes an American newspaper, *The Sunday Sentinel*, advising mothers in 1914 to 'use pink for the boy and blue for the girl, if you are a follower of convention'. According to Wikipedia, 'The first inauguration of Dwight D. Eisenhower (1953), when Eisenhower's wife Mamie Eisenhower wore a pink dress as her inaugural gown, is thought to have been a key turning point in the association of pink as a colour associated with girls.'[7]

There is, of course, scientific evidence offered in favour of the more traditional view of gender as the unwinding of a programme inscribed in the sex chromosomes. In an article in *The Guardian* entitled 'Male and Female Brains Wired Differently, Scans Reveal',[8] readers were treated to pictures of brightly coloured brains, highlighting the distinctive pathways in brains belonging to male and female subjects. According to the article, 'Ragini Verma, a researcher at the University of Pennsylvania, said the greatest surprise was

---

[6] <http://www.jeongmeeyoon.com/aw_pinkblue.htm>
[7] <https://en.wikipedia.org/wiki/Pink>
[8] <https://www.theguardian.com/science/2013/dec/02/men-women-brains-wired-differen tly>. Referring to Ingalhalikar et al. 2014.

how much the findings supported old stereotypes, with men's brains apparently wired more for perception and co-ordinated actions, and women's for social skills and memory, making them better equipped for multitasking.' The only comment required on this, as on a great deal of the research in this area, is: Why assume the wiring of the brain is a cause rather than an effect of gender difference? It would be extraordinary if decades of different experience did not result in significant differences in the brain. Avoidance of the confusion of causation and correlation is the first methodological rule taught to every neophyte scientist.

The point I want to make with this last example is not to ridicule the work of a respected group of scientists, but to note how powerful an unquestioned framework can be in shaping a research project and interpreting its results. Thinking in terms of the mechanistic unfolding of a genetic programme—gene sequence causes brain structure causes behaviour—it is natural to assume that brain structure is caused by sex-specific genes and in turn causes the stereotypical gendered behaviour of men and women. But thinking in terms of a developmental process, driven by multiple factors, then it is the *surprise* that the brain 'supported old stereotypes' that is bizarre. Of course people are trained to conform to normative stereotypes; and of course their brains come to manifest structures that serve this conformity.

Sex/gender, then, is properly seen as what I have called a homeorhetic process, a process stabilized, often by a wide range of factors, to follow a particular path. This is a good point to recall Waddington's iconic representation of a homeorhetic process (chapter 4, figure 4.1). The widest canyon leads to the 'normal', that is, most typical, outcome. But changes to the stabilizing conditions—see the pegs under the landscape—may change the topology so as to divert the ball down a side canyon. Or, in fact, merely stochastic movements (from the point of view of the model) of the ball itself may result in diversion down a different pathway. The stabilizing factors, as just mentioned, include such things as hormones, parenting, schooling, social environment. Some of these, no doubt, are affected by genes, and these may change the shape of the landscape. Recall also that there are different starting points, including genotypes different from XX and XY (such as X0, XXY, XYY). One might think of these, and perhaps many other genes on autosomes (chromosomes other than those that are sexually differentiated), as imparting a different spin or velocity to the starting ball, thereby changing the probability of diversion onto a canyon other than the main one.

So what does this tell us about the empirical work on sex/gender, ranging from the kind of neuroscience just mentioned to the vast quantities of empirical investigation, mainly by Evolutionary Psychologists, of such questions as what body shapes people find attractive (Singh 1993) and whether they are willing to have a date with an attractive stranger who propositions them on the street (spoiler: men often will; women never) (Clarke and Hatfield 1989). The first thing to say about all such research is that it provides a snapshot of a process, not an insight into the essence of a thing. Such research, generally carried out on American college students, describes a particular stage in the development of this human cohort. From this point of view, the fact that women who have grown up in a society in which sexual violence is an everyday fact of life are unwilling to have a potentially intimate relation with an unknown man is hardly surprising. If anyone wants to object that violence against women is just another fixed outcome of the genetic programme in men, they should look at the variability of relevant crime statistics. Although it is difficult to get fully accurate statistics on such matters, the variability is hard to dispute. The USA, for example, is reported to have twenty-seven times the per capita incidence of rape of Japan,[9] and these are not even the extreme outliers.

Tempting though it is to dodge this heated and difficult issue entirely, I should say something about the implications of all this for transgender, though I shall try to be as bland and uncontroversial as possible. The first and simplest thing to say is that this is clearly one outcome of the development of sex/gender, one canyon in the Waddington diagram. Second, given this, no one should be condemned or mistreated for ending up with this outcome. Whether this pathway is encoded in the genes or develops in childhood, for most transgender people their condition is clearly no more experienced as an optional choice than is homosexuality. Surely everyone should have the right to decide how they dress, what name they prefer, or the pronouns with which they would like to be identified.

I shall not hope to offer a solution to the most fraught questions of all, concerning who should be allowed in which toilets, gyms, dressing rooms, or prisons. I note only that the developmental processes I am discussing are not simple and the outcomes do not fall into homogeneous kinds. In deciding who should be allowed to use, for instance, women's shelters, the answer must surely lie somewhere between, on the one hand, anyone who says they are a woman, and on the other, only people who were identified as female at birth. Where on this spectrum the line should be drawn could at least be

---

[9] <https://www.nationmaster.com/country-info/compare/Japan/United-States/Crime>

acknowledged to be a matter about which well-meaning people might disagree and enter into serious discussion.

Finally, there is a question of language. Woke-bating politicians like to ask whether transwomen are women, or to define women as menstruators, possessors of wombs, etc. It is, I think, unhelpful to deny that there is a biological frame within which 'woman' has been used to mean female human, and 'female' to refer to a member of the sex that produces larger gametes. Why this should be the only acceptable usage in a population in which most people have never heard of a gamete and still less of the significance of relative gamete size, is another matter. Gender and identity are typically more salient aspects of a person's life than the biological details of reproduction.

The terms sex and gender were introduced by second wave feminists to insist that the social and behavioural differentiation of men and women (gender) was grounded in society not biology (sex). For several reasons, including the question of transgender, the distinction eventually proved simplistic. The usage 'sex/gender' is intended to point to the difficulty of disentangling these two concepts. Not all women, even defined in terms of gamete size or chromosome type, menstruate or have uteruses. One obvious thing to say is that there are many kinds of women, of which transwomen are one. But surely this doesn't mean we always need to qualify the use of woman by parenthetically adding the kind to which a particular woman belongs. This would be much like supposing we always needed to mention someone's race whenever referring to them. Most of the time this would be irrelevant and offensive.

Race

So let me now turn to the other most widely discussed systematic variation between humans, that which is referred to by race. In the nineteenth century there was a heated debate over whether distinct human races had evolved independently on the different continents or whether they were variations within a single species. The latter argument was decisively successful, and now even the most unhinged racists are unlikely to claim that different human races are distinct species or even subspecies.

Debate over how exactly races should be understood remains alive and well, however. Partly this reflects a good deal of more general uncertainty about divisions within a species. The only formally acknowledged taxonomic rank below the species for animals—plants and fungi are more complicated—is the subspecies, though even this is not very clearly defined. Some degree of physiological and genetic distinction, and some kind of isolation, usually geographical but perhaps merely behavioural, is required to qualify as a subspecies, as subspecies are by definition members of the same species and could, therefore, interbreed if there were no barriers to their doing so. The human species is generally acknowledged to be too young and mobile to have evolved subspecies. Too young, because populations need to be isolated for a considerable time to evolve the sort of differences that might qualify for subspecific distinction, and too mobile, because human migration has quite quickly broken down temporary isolation. The scientific consensus is that *Homo sapiens* is a very widespread but rather genetically homogeneous species, with some local variation, largely reflecting adaptation to local climate. Physical anthropologist Nina Jablonski provides compelling evidence that skin colour is a highly plastic, that is, rapidly evolving, adaptation to balance vitamin D production and ultraviolet damage in different regimes of solar irradiation (Jablonski and Chaplin 2000, 2017).

Whatever the scientific consensus, many or most people take the evidence of their eyes to show that there are distinct races of people, and they believe that they can readily tell to what race a person belongs, at least in most instances. For a long time the scientifically informed public were, nonetheless, willing to accept what the scientists told them. This was no doubt in part because the grim history of racial discrimination made the non-existence of race an attractive finding, but also because advocates of this view had persuasive and understandable ways of making their point. In particular, research by the distinguished population geneticist Richard Lewontin showed that for any two humans, the genetic differences attributable to race were very small compared to the typical differences between people of the same race (Lewontin 1972). Thus any given individual could have more genetic similarities to a person of a different race than to one of their own race. As Lewontin wrote:

> based on randomly chosen genetic differences, human races and populations are remarkably similar to each other, with the largest part by far of human variation being accounted for by the differences between individuals...(1972, 397)

He concluded that:

> Human racial classification is of no social value and is positively destructive of social and human relations. (1972, 397)

This argument has itself been widely debated, but on the whole it has re-inforced the consensus: human race is biologically unimportant and largely a social construct. Indeed, echoing the critique of race science as a justification of colonialism, it was increasingly argued that race was a consequence of racism rather than racism a response to race. However, the rise of genomic science has led some to think that the scientific status of racial categories needs to be rethought.

Here is a quote from the sometimes controversial but well-known science journalist Nicholas Wade:

Analysis of genomes from around the world establishes that there is a biological basis for race, despite the official statements to the contrary of leading social science organizations. An illustration of the point is the fact that with mixed race populations, such as African Americans, geneticists can now track along an individual's genome, and assign each segment to an African or European ancestor, an exercise that would be impossible if race did not have some basis in biological reality. (Nicholas Wade, *Time*, 9 May 2014)

The basis of this claim is a widely cited and discussed 2002 paper in the journal *Science*, by Noah Rosenberg and colleagues, 'Genetic Structure of Human Populations'. Rosenberg studied several hundred genetic loci from over a thousand individuals belonging to fifty-two populations. The loci they used were microsatellites, repetitive elements of the genome generally thought to be non-functional and therefore subject to high rates of mutation and minimal selection. Their research confirmed Lewontin's much earlier finding that at most 5 per cent of the overall variation was between major groups as opposed to between individuals within groups. Nonetheless, they also were able to use these genetic data to sort individuals with a high degree of accuracy into groups that corresponded well with geographic regions and familiar racial categories. This the basis of Wade's claim.

But what does this research really show? A crucial point to note concerns the programme, called *structure*, that was used to analyse these genetic data. Using *structure*, it is necessary first to specify the number of groups, K, into which the data are to be sorted. Having done this, the programme selects the best clustering of the data into this number of groups. Starting with K = 2, the cluster identifies groups centred on Africa and America. Each unitary increase in K divides one of these groups, and by K = 5, it divides the data into familiar major geographic regions of origin: African, Asian, European, American, and Oceanian. K = 6, however, splits off the isolated Kalash group

in north-west Pakistan, a unique and persecuted group with a religion related to Hinduism and a population of about 4,000. As is well understood,
small, isolated groups may be subject to rapid genetic drift, so it is no surprise
that such groups begin to appear as K is increased.

So what are we really seeing here? The human lineage is a process, let us
recall, and for the last 100,000 years it has been an extremely dynamic one,
migrating from East Africa across the globe and, incidentally, wiping out
all other species of humans along the way, whether by violence, competition, or just absorption. If we think of the migration pattern as a tree then,
very roughly, bearing in mind cases such as the Kalash, the further from a
common ancestor in the tree the greater the genetic diversity through drift.
So *structure* is, more or less, mapping the migrational history of *Homo sapiens*. Recall that we have been considering a genetic measure of difference
that is explicitly chosen to exclude or minimize selective evolution. But no
doubt these populations have been subject to some selective evolution in
terms of superficial phenotypic differences, as well as experiencing some
genetic drift. Thus, we can observe phenotypic differences that give us a
fairly reliable basis for inferring ancestral history. As I mentioned, skin
colour is a trait particularly susceptible to rapid locally adaptive evolution.

But the tree generated by structure does not give us an evolutionary history marking lineage divides. We are looking at processes within the lineage,
and we are tracking dominant movements, not unidirectional movements.
We should not imagine a committee 50,000 years ago deciding that it was
time for everyone to move to China. Some people are moving against these
dominant flows, and if there is selection for traits that are not merely of
local value, they are likely to spread into other branches of the tree. The
genetic information, in short, is not about different kinds of people, it is
about the ancestries of people in terms of flows within the wider human
lineage. And contrary to the idea that showing different racial (or ancestral)
groups have detectable genetic differences supports the hypothesis of racial
difference, the genes in question are, again, chosen for their lack of phenotypic or physiological significance. The notion that they provide support
for the expectation of systematic differences in behaviour, or behavioural
capacity, is not only lacking in any biological rationale, but it also carries
the serious risk of occluding the differences in social and economic background, opportunity, and experience that surely do cause difference in the
traits of people whose lives are shaped by racial classification. This is where
such research, unless very carefully interpreted, can be seriously harmful.

To conclude this discussion, biological race in humans does not mark a phenomenon of any biological importance, it is rather the reified legacy of a historical process, the migratory history of our species. In fact, the research by Rosenberg et al. shows the very opposite of what Wade suggests—and indeed a claim very similar to Lewontin's on the far greater variation within than between races is stated in the Rosenberg paper and, even, at one point in Wade's article. We should not think of biological races as defining different kinds of people; rather they are the genetic and phenotypic souvenirs that individual humans or groups of humans have acquired on their particular path through the complex process that is the recent human lineage.

Race is biologically trivial but remains detectable both genetically and phenotypically. Why? The human lineage contains a huge mélange of migrations large and small, but most of these are lost in the melting pot of human reproduction. At least one of my ancestors on both sides of my family, I believe, was a Huguenot, part of a major migration from France to various parts of Europe in the late seventeenth century, and one of them bequeathed me my last name. Subsequent migrations established Huguenot settlements in North America, South Africa, and Australia. No doubt they took some genetic markers with them, and there is a chance that I carry some of these markers. And certainly, there were skills and practices that were dispersed with the people, some of which survive today. But recalling two of my central theses, that a process requires active stabilization and a lineage is a process, apart from a few isolated societies and churches in the USA, there are no institutions that maintain the Huguenot cultural lineage. Most descendants of Huguenots, like me, maintain this heritage as no more than a historical curiosity.

Why are races so much more persistent? Precisely because they are sustained by robust cultural processes. This is the deep truth, I suggest, to the immediately paradoxical claim that racism causes race rather than race causing racism. The normal fate of a migration within the geographic range of a species, human or otherwise, is dispersal among the current inhabitants. What prevents this is creation of barriers to intermingling and especially to interbreeding. This is most commonly achieved by the stigmatization of immigrants as belonging to inferior races or inferior cultures. We should, I believe, hope for the end of race as anything more than the historical curiosity that I might claim for my trace of Huguenot ancestry. This is not, of course, to deny that there are pressing questions of justice to be addressed before it will even be morally desirable to forget the concept of race. But I do insist, following Lewontin, that our goal should be the end of any racial classifications.

There may be a concern that the end of race would entail a loss of cultural diversity. The simple response is that there are surely no race-specific cultures. Even if we exclude such phenotypically black people as native Australians, and limit the word 'black' to people with African ancestors from sub-Saharan Africa within the last few centuries, there is no reason to expect a common culture. It is revealing that the idea of a white culture suggests only racist ideology. The kinds of cultures we might want to preserve are much narrower: Zulu or African-American; Basque or Italian.

Would the end of racial distinctions in the US mean the end of Jazz, for instance? Obviously not. Though Jazz has origins in Africa and then in black populations in America, there are now plenty of white or Asian jazz musicians. Cultural traditions are maintained by many practices apart from racial distinction, in this case by the training and performance of a large community of musicians.

Cultural Kinds

Let me conclude with some brief reflections on cultural lineages. There is no doubt that all humans have a place in both the human, biological lineage with its many flows of people and their biological idiosyncrasies, and many cultural lineages. Cultural lineages include skills, sports, art forms, nationalities, religions, languages, and much more. All of these are processes, sustained, if they are sustained, by the practices of their members and adherents. Some, like membership of large nation-states, can function rather like races. They are more or less pure vehicles of distinction maintained not by the particular characteristics of their practitioners, but by the institutional imposition of borders, passports, voting rights, and suchlike. These are reified into different kinds of things for various, sometimes questionable, purposes by enthusiastic nationalists. While there are many things I admire that are more or less strongly associated with Britain or with England, the nationalities I still carry since my European identity has been taken from me, I don't think they depend on these institutions. Pubs, British universities, cricket, socialized medicine free at the point of delivery, the BBC, the Notting Hill carnival, Burns night, Christmas dinner, scotch whisky, and on and on, are not dependent on the maintenance of borders. On the contrary, openness to the cultural traditions from around the world and the people who bring them has constantly enriched the lives of British people. Indian and Chinese food among countless world cuisines, rock music and grand opera, world literatures, this list is even longer than the last.

The richness of the human process at its best is its ability to differentiate into communities that generate new possibilities for human life and practice and, often, to disperse these around the human species. On the whole, these are maintained not negatively, as is the case for biological lineages such as race that persist only by deliberate isolation from the surrounding genetic flow, but positively, by the enactment of the practices that constitute them. It may be necessary to protect some of the smaller cultural flows—languages, religions, and traditions of small communities—with more boundary-like measures to prevent their diffusion into the dominant culture, but this is the preserving of diversity as a resource rather than its exclusion as a threat. Some kind of isolation is needed for cultural preservation and cultural evolution, but usually this is sufficiently provided by the discipline and dedication of adherents or practitioners. A great deal of human diversity, and much that is to be treasured, can be understood in terms of the location of the individual in these various cultural lineages.

But, finally, the human species has reached a point where a further perspective is essential, one that cuts across all these local cultures to the trajectory of the species as a whole and the niche—more or less the world—that it inhabits. We are a species currently in a process of deconstructing its niche. Our eating practices, our individualized and carbon-intensive modes of travel, our general overconsumption, our ideological commitment to economic growth, and more are changing the entire planet in ways that threaten its ability to support our own lives.

The solution to this problem and no doubt others is increasingly hampered by competitive ideologies at both the individual and national level and the boundaries that continue to frame these competitive principles. We need to instil in people the sense of being human, part of the human species and not just as being an instance of the human kind, but as being part of the process which is the human species.

We might learn something from Native American thought here. The Seventh Generation Principle is based on an ancient Haudenosaunee (Iroquois) philosophy that the decisions we make today should result in a sustainable world seven generations into the future. This is reflected in a powerful statement by Rick Hill Sr, a citizen of the Beaver Clan of the

Tuscarora Nation of the Haudenosaunee at Grand River, painter, carver, photographer, basket weaver, and consultant to National Museum of the American Indian, Smithsonian Institution:

> If you ask me what is the most important thing that I have learned about being a Haudenosaunee, it's the idea that we are connected to a community, but a community that transcends time.
>
> We're connected to the first Indians who walked on this earth, the very first ones, however long ago that was. But we're also connected to those Indians who aren't even born yet, who are going to walk this earth . . . And our job in the middle is to bridge that gap. You take the inheritance from the past, you add to it, your ideas and your thinking, and you bundle it up and shoot it to the future. You come here to work hard so that the future can enjoy that benefit.[10]

Here I see ethical thinking grounded in the assumption that we are part of a process of which we are merely a transient part. Could we even imagine extending such thinking to an awareness of being part of a species, or even perhaps of a vastly larger process still, of life on Earth? Again, this is not a thought that should be grounded in membership of a kind, sharing of an essence, but in being part of a process. It seems increasingly clear, as the degradation of our planet and the drastic changes to our climate resulting from human activity increase by the day, that unless we can achieve and take seriously such a perspective, the future of the human species and much else of life on our planet is in grave peril. Failure to fit one's own interests to the limits imposed by the importance of the health of this larger process is, almost literally, cancerous. This way of thinking may sound an ambitious and unlikely achievement, but perhaps it is the only way that human civilization and perhaps even human life can be sustained for much longer.

[10] https://www.pbs.org/warrior/content/timeline/opendoor/roleOfChief.html#:~:text=%22If%20you%20ask%20me%20what,a%20community%20that%20transcends%20time.

6

Free Will

Introduction

Throughout these chapters I have emphasized the deeply social nature of the human species. This may seem inevitably to devalue the human individual. Paradoxically, however, I want to argue that the reverse is the case. It is in large part the social character of the human that makes possible the autonomy of the human individual. In this chapter I shall try to explain.

Mainstream philosophy has an almost contradictory set of views about the individual. In much political philosophy the autonomous choice of the individual agent is paramount. In political theory democratic processes aggregate individual decisions. In economic theory individual preferences and choices determine the production, distribution, and prices of products required to meet human needs and wishes. On the other hand, for centuries mainstream metaphysics has denied human choice.[1] The doctrine of determinism has been taken to deny the efficacy of the human choices so central in political and economic theory. I admit that this is a contentious description of the situation, so I shall begin this chapter by outlining the debates over free will and determinism, and attempting to justify my description. I shall then begin to explain how the process philosophy I have been advocating provides a path out of the problem.

Many philosophers, even today, are determinists. The view is less prevalent, I am pleased to say, among philosophers of science. Many, however, continue to hold that exceptionless physical laws determine everything that happens. When reminded that physics nowadays has eschewed determinism in the theory of quantum mechanics, they will reply that all

---

[1] The debate on free will has generated a huge amount of discussion among philosophers. I cannot attempt here to give adequate citations to this literature. As with many philosophical topics, an excellent place to enter the debate and find numerous further citations is the article in the online *Stanford Encyclopedia of Philosophy* (O'Connor and Franklin 2022).

Everyone Flows. John Dupre, Oxford University Press. © John Dupré 2025.
DOI: 10.1093/9780198941866.003.0006

the indeterminism disappears at larger scales.[2] Schrödinger's cat is an intriguing thought experiment, but in the real macroscopic world universal law holds sway.

I don't propose to discuss in any detail how this curious belief came about as it should be clear from my chapters to date that I have a very different view of the world. I cannot resist, however, quoting an exceptionally insightful comment from the late Cambridge philosopher Elizabeth Anscombe:

> The high success of Newton's astronomy was in one way an intellectual disaster: it produced an illusion from which we tend still to suffer. This illusion was created by the circumstance that Newton's mechanics had a good model in the solar system. For this gave the impression that we had here an ideal of scientific explanation; whereas the truth was, it was mere obligingness on the part of the solar system, by having had so peaceful a history in recorded time, to provide such a model. (Anscombe 1971)

As a therapy against this illusion, I might also recommend Liu Cixin's (2015) wonderful science fiction novel, *The Three-Body Problem*, which, among other things, explores the nature of life on a planet under the chaotic gravitational influence of three suns.

At any rate, for determinists it is liable to appear that individual choice is an illusion. The nexus of physical laws determines what happens now, which in turn determines what happens in the future. Similarly, what happens now was preordained by events that happened long ago. My belief that I choose between the apple and the orange, or hard physical labour and a life of crime, is just that, an illusion. For these decisions were determined long ago, perhaps at the moment of the big bang.

Since David Hume's famous thoughts on this topic the dominant view has been to embrace this conclusion with equanimity. Hume distinguished between liberty of spontaneity and liberty of indifference. What we imagine we want is the latter, the possibility of acting in either of two ways. But why should we want this? When I decide that, all things considered, I prefer the apple, why is it important that I somehow be able to choose the orange? What I should care about is just that my choice reflects my authentic preference, something available, he writes 'to every one, who is not a prisoner and in chains' (Hume 1772/1993, I. 8. 73). One might extend this list to further

---

[2] Some philosophers of physics have recently claimed, contrary to common opinion, that quantum mechanics is in fact a strongly deterministic theory (Chen 2023).

sources of unfreedom such as coercion by threats, psychological manipulation, or insanity. But absent some such factor we are all free to act on our (determined) preferences. This has become the dominant opinion of contemporary philosophers.

In defence of this view, it is often pointed out how little we should care about liberty of indifference. Why should I care about being able to get what I don't want? Another way of thinking about indifference is to imagine that our decisions have an element of randomness to them. But this sounds nothing at all like what I might want if I want to be free. The compatibilists, as followers of Hume's lead are generally known, insist that freedom is precisely the confidence that I will be able to do what I most want to do, something that would be clearly frustrated by an element of randomness.

Perhaps it remains unsurprising that the enemy of compatibilism, generally known as the libertarian or voluntarist, remains unsatisfied. What is missing is a sense of actual influence or power. I imagine that my actions make a difference in the world; not merely as a conduit for the long ago determined causal nexus to flow through, but as an actual source of change. This is sometimes connected to the idea of 'agent causation', according to which the agent is the originator of a causal chain or process (see e.g. Clarke 1993). To a philosopher immersed in Hume's causal world, in which causality is precisely the enactment of causal regularity, agent causation is liable to seem distinctly mysterious. Either the agent is located somehow outside the normal causal order, as in a Cartesian soul, or they are an unexplained anomaly in an otherwise well-ordered world. Neither alternative seems attractive. I shall suggest, however, that process philosophy offers a way forward.

Let me again stress how very different the world I want to describe is from the determinist's. As opposed to everything being ordered by universal lawful regularity, I see a background of chaos. Order is something that emerges occasionally under special circumstances and that lasts for a finite period of time. Whereas the voluntarist philosopher typically tries to see the human as an exception of some kind to an otherwise all-encompassing causal order, I see the human as the pinnacle of order, the most intricately stabilized system yet known to have emerged, or evolved, from the surrounding chaos.

Rather than laws, which have widely been recognized as increasingly hard to find, at least in biology, I prefer to think in terms of causal powers (Mumford and Anjum 2011), capacities to affect the world in systematic ways that have emerged as complex systems stabilized themselves. Here again

I must make a nod to Hume, who is also famous for convincing many philosophers that causal powers were nonsensical, lacking in any possible empirical grounding. Fortunately, though this debate still rages, there is a solid anti-Humean tradition with which I can align myself. I will again single out Elizabeth Anscombe. Hume, she notes, challenges us to 'produce some instance, wherein the efficacy [of a causal power] is plainly discoverable to the mind, and its operations obvious to our consciousness or sensation. Nothing easier: is cutting, is drinking, is purring not "efficacy"?' (Anscombe 1971).

Once again, we see the power of a framework to determine what we see. I see the rain washing the dirt off my car; I see the dog gnawing on a bone; I see the boy catching a ball; everywhere entities (processes) interact and bring about changes in one another. But suppose I think the world I see is a world composed wholly of discrete things. Then I see a ball moving through the air, then I see it reaching the boundaries of the boy's hand, then I see that it no longer moves. Do I see the boy's hand *making* the ball stop.[3] Surely this is not something I see in addition to what I just described? So perhaps as Hume would say, it is something added by my mind.

Suppose instead I see the world as full of activity, process. I see these processes intersecting—the rain and my dirty car, the dog and the bone, the boy and the ball—and affecting one another. This, I suggest, is just what causation is, and we see it all the time (Dupré 2021b). This is why, as Anscombe's response illustrates, our descriptive language is laden with causality. Our constructed world is full of things we describe in terms of their causal powers—knives cut, tables support, radiators warm, ovens cook, and so on—and adjectives that qualify these powers—sharp knives cut more easily, wobbly tables support unreliably, airlocked radiators fail to warm. Unsurprisingly, though, it is verbs, action words, that are the most process filled. Verbs describe processes—thinking, walking, boiling, singing—and causal interactions—touching, throwing, lifting, digging, breaking, pushing, and countless others. These activities are powers that the entities we observe have, and powers we sometimes observe them exercising. In sum, then, unless we adopt an ontological frame that excludes the observation of the exercise of powers, this is something that we observe constantly and that permeates all our descriptions of the world.

---

[3] Since Albert Michotte's work in the 1940s, it has been widely accepted by cognitive psychology that we *do* see the world as structured by causal relations.

Mechanism and Process

So, in this chaotic world, we have entities with causal powers. Again, we can contrast the two world views. In the traditional metaphysics, the equivalents of casual powers are the lawlike regularities to which things are subject. If a thing is complex, these regularities are consequent on the regularities that apply to its components. This gives rise to the idea of a machine, the parts of which are organized so as to bring about a regular behaviour: when I press the switch the blades in the food processor rotate rapidly, for instance. Most of the time the machine just waits patiently for the switch to be pressed.

Unlike the machine, a process acts constantly to maintain itself and its structure. Whether or not the structure explains its behaviour, like the machine, the behaviour also explains the structure. Evolution has generated processes that are extremely good at maintaining their structure, and we need to look at the kinds of causal powers that have enabled them to do this. While I have been sceptical about the kind of openness of behaviour suggested by Hume's liberty of indifference, it does point to something that might help understand the problem of free will, which is the indeterminacy of outcome. This is something that is almost always seen as a failing in a machine, but something of this sort seems to be implicated in our conception of freedom.

Indeterminacy is not, of course, the same as variability. I can imagine a machine that identifies animals. When it is in the vicinity of a rabbit, it types 'rabbit'. When a fox approaches, it types 'fox'. Its behaviour is variable but, we hope, determinate. We hope that it will always type the same words when confronted by a particular type of animal. Much animal behaviour is often interpreted in this way.

A central paradigm of mechanical animal behaviour is the Sphex wasp, observations of which date from Jean-Henri Fabre in the nineteenth century, but which has become a standby of discussion in philosophy of science and cognitive science especially due to the work of Daniel Dennett (1973). This wasp, like many other related species, provisions a nest for its larvae with the paralysed body of another insect, in this case a cricket. The sphex typically brings its prey to the entrance of a prepared burrow, checks the state of the burrow, and then comes back up to drag in the victim. Here is Fabre's account of his famous observation of the wasp, *Sphex flavipennis*:

> [A]t the moment when the Sphex is making her domiciliary visit, I take the
> Cricket left at the entrance to the dwelling and place her a few inches far-
> ther away. The Sphex comes up, utters her usual cry, looks here and there in

astonishment and, seeing the game too far off, comes out of her hole to seize it and bring it back to its right place. Having done this, she goes down again, but alone. I play the same trick upon her; and the Sphex has the same disappointment on her arrival at the entrance. The victim is once more dragged back to the edge of the hole, but the Wasp always goes down alone; and this goes on as long as my patience is not exhausted. Time after time, forty times over, did I repeat the same experiment on the same Wasp; her persistence vanquished mine and her tactics never varied. (Fabre 1915/2001, 43)

It is concluded that sphex's behaviour is a wholly mechanical routine: place cricket at entrance to the nest, check condition of nest, drag cricket in. If the routine is disrupted the wasp simply tries again to execute it. This story has been cited repeatedly in support of mechanistic accounts of animal behaviour.

Unfortunately, as is shown in a rather more thorough examination of the history of research on these wasps by Dutch philosopher Fred Keijzer, the example is poorly chosen. Not only subsequent research, but even research by Fabre himself, shows that this observation is far from universally true of the wasp. A few lines below the text just quoted, he writes:

Having demonstrated the same inflexible obstinacy which I have just described in the case of all the Sphex-wasps on whom I cared to experiment in the same colony, I continued to worry my head over it for some time. What I asked myself was this: 'Does the insect obey a fatal tendency, which no circumstance can ever modify? Are its actions all performed by rule; and has it no power of acquiring the least experience on its own account?' [...] good fortune brought me into the presence of another colony of Sphex-wasps, in a district at some distance from the first. I recommenced my attempts. After two or three experiments with results similar to those which I had so often obtained, the Sphex got astride of the Cricket, seized him with her mandibles by the antennae and at once dragged him into the burrow. [...] At the other holes, her neighbours likewise, one sooner, another later, discovered my treachery and entered the dwelling with the game, instead of persisting in abandoning it on the threshold to seize it afterwards. (Quoted in Keijzer 2013)

Keijzer concludes that the sphex teaches us more about philosophers and cognitive scientists than about insects. A model of an entirely mechanistic insect can provide either a contrast with our own condition or an illustration of that condition depending on the taste of the researcher. What Keijzer shows, however, is that while sphex may not be the smartest creature around, it is often able to see what it is doing and improve its route to the goal. This leads us to a slightly different version of the contrast between mechanism and process, that between mechanism and agent.

Agency

Let me restate the problem of free will from an explicitly mechanistic perspective. A mechanism, when it is working, is a system in which the properties of the part determine the behaviour of the whole. So, if I am a machine, either the properties of my parts cause me to act a certain way, or they don't. In the first case I am caused to act merely because of the arrangement of my parts, so I am not free. In the latter case, either chance, or something external to me, determines my behaviour. So again, I am not free.

A process is not like this at all. It has changing properties and it is of the nature of many processes, notably living ones, to change the world around them. Think of the damage done by a paradigmatic process, a storm. A machine can be used to do damage, of course, but it is not of its nature to do so as it is of a storm. Left to itself it will generally do nothing. An organism is a great deal more complicated and organized than a storm. But in a consequently more targeted way, all organisms constantly change their environments. We might compare the damage done by the storm to that inflicted by an elephant or a tiger on, respectively, flora and fauna.

But though we may speak metaphorically of a storm as an agent of change, no one seriously thinks a storm is an agent. The status of tigers and elephants in this regard is more contentious. What I want to suggest is that we think of agency as pertaining to some degree to anything that has goals and acts to further them. A storm certainly has no goals. More debatable, perhaps, is the case of a thermostat. But I don't think a thermostat has goals or interests; though it has a tendency to achieve a certain state, this is not its goal; it serves the interests of the people who made or use it. Similarly, though an immune system can respond in very subtle ways to the contingencies of its environment, it has no goals distinct from that of the organism of which it is part. But we might think of a very simple organism, such as a bacterium swimming in the direction of a chemical gradient towards an environment more favourable to its survival, as an agent. All organisms, one might say, have an interest in survival and act in ways that promote that interest. Concerns about the future of robots, or even computer programs, often centre on the possibility that they might acquire goals of their own, goals potentially in conflict with those of their makers. They might, that is to say, become agents.

Living systems are typically opportunistic. All, or almost all, have the capacity to respond to the unpredictable contingencies of a changing environment in ways that may maintain their stability (survival). This is a capacity that one should expect to be favoured by natural selection. It is possible that the best way to deal with the world is mechanistically, as Dennett imagined the sphex to do. But, as the smarter sphexes quickly illustrate, it is possible to do better and organisms generally do. If entomologists became a common enough nuisance to the lifeworld of sphexes, it is likely that they would soon evolve to become smarter.

Actually, it appears that even the more repetitive sphexes may not be so dumb. Keijzer reports a number of more recent studies on the behaviour of similar species, concluding with a very extensive study over five years by Jane Brockmann (1985) on the sphex species *Sphex ichneumonius*. On the basis of this work she concludes that: 'Where responses show stereotypy, such as in repeated prey retrievals, there is an obvious, adaptive explanation.' Specifically, she notes, 'The fixity of repeatedly repositioning and reentering the nest is almost certainly an adaptive response to prey that can easily become lodged in the nest if pulled in backwards' (Brockmann 1985, quoted in Keijzer 2013). In some, probably rare, cases, behavioural rigidity is adaptive, and systems may develop in this mechanistic way. But this is a bad model for biological systems generally (perhaps rather as the solar system is a bad model for dynamical systems generally). When plasticity is adaptive, systems will develop flexible, minimally agentive, ways of interacting with their environment. And it seems, both empirically and as a matter of common sense, that plasticity is very frequently adaptive. As Keijzer goes on to remark, the moral of research on these insects seems to be the opposite of that usually drawn from it: where there is a benefit to it, animals have generally adopted flexible behaviour.

So many organisms, perhaps all, exhibit a kind of agency, at least in the fairly undemanding sense that I have given the term. But perhaps we should make the concept a bit more demanding. The sphex, after all, even given Dennett's conception of it—let's call it the D-sphex—responds adaptively to its environment. When it comes out of its nest and finds the grasshopper where it should be, it pulls it down into the nest. When it sees it some way away it pulls it back to the nest. Both of these are adaptive responses to the environment. In the second case it then makes a pointless trip down into the nest. But no one is perfect.

The problem with the D-sphex is that its behaviour is entirely determined by its environment. Whether its behaviour serves its goals is irrelevant. Of course, if it is lucky, it will have evolved so that the deterministic response to the environment is generally a good one. But that is no part of why it acts that way. That is solely to be explained by the lucky behaviour of its ancestors that enabled them to be ancestors. But the R-sphex, R for real, is not so simple. When its circumstances change it is sometimes able to change its behaviour in ways that are more appropriate to its goals. Something like this provides a more demanding but plausible condition for (minimal) agency.

Am I saying that the R-sphex has free will? Let me sidestep that question for a moment, and just look a bit further into agentiality. I have said that being an agent is a matter of degree. A thermostat has none or almost none, a bacterium has a very little, and a human has a lot. What are the variables that contribute to greater agentiality?

First, agents have goals of some kind. They have preferences about how the world should be, and their actions have some tendency to make the world that way. Subject to some doubts about the nature of preferences, this criterion might even attribute some minimal agency to thermostats.

Second, agents learn. Their capacity to respond to the environment is a function of preceding experience with similar environmental contingencies. This is a permissive criterion, and it is beginning to seem that it may well apply to a great many organisms. Learning is of course most studied in animals, but a number of scientists now believe that plants can learn from experience, though this remains controversial (e.g. Calvo et al. 2020). Artificial intelligence systems also certainly qualify, reinforcing the unsurprising point that artificial intelligence has some features in common with natural intelligence. More traditional machines—pumps, toasters, lawnmowers, etc.—surely don't learn, though in a minimal sense computers surely do. I am the most salient part of the environment for my laptop, and on the basis of its interactions with me it is able to show files, run programs that I have installed, and so on. Since an increasing proportion of machines now contain computers, I suppose that many of these are acquiring minimal capacities to learn.

Perhaps a more interesting point is that an evolving lineage might claim to satisfy this criterion of agentiality, on the basis of learning by natural selection. Something like this might justify the claim that bacteria learn, for example to combat antimicrobial chemicals. It is not generally suggested that individual bacteria learn how to do this—though one might perhaps think that they could learn from their neighbours through the process of lateral gene transfer. But the rapid generation times of bacteria colonies, together with the genetic resources that may be distributed through a local population, allow evolution by natural selection to provide a rapid response to a novel environmental challenge within a bacterial lineage. This also connects with the debate whether the individual bacterial cell, or rather the colony of many cells, is more appropriately seen as the organism.

Third, agents exhibit some degree of creativity. Again, I do not want to make this concept very demanding. Indeed, it might be that it is no more than any behaviour that is not a purely instinctive and deterministic response to a stimulus. But I do want to add some level of appropriateness that distinguishes creative behaviour from the merely random. The R-sphex is being creative when it brings the cricket back from where the experimenter has put it and drags it into its nest. It would, at any rate, be a different kind of response if it had, say, dragged it another metre from the nest, tossed it in the air, and abandoned it. It might be the case, I suppose, that creativity consisted, for some creatures, merely in generating random actions and observing whether any of them were conducive to its goals. But this seems a rather bad way to act and immediately suggests a fourth and more ambitious feature of agentiality, perhaps well-developed only in animals with reasonably complex nervous systems, foresight, and, ultimately, intelligence.

Some agents are able to imagine the consequences of novel actions drawing on experience or (in the human case, at least) socially transmitted knowledge. They thus have some insight into the likely value of novel responses to contingencies. Combined with the ability to learn from the outcomes of novel behaviour we are getting close to a rudimentary experimental method. An animal that can discern potentially valuable forms of behaviour and evaluate their results has an extraordinary capacity.[4] Such foresight is certainly not unique to humans, though it seems rare. New Caledonian crows, for instance, have become famous for their tool-using abilities, and seem able to see several steps ahead in devising solutions to complex problems.

I don't much care whether this is an account of agentiality that fits well with standard appeals to the concept, or even, in the present context, whether

---

[4] This brings to mind Dennett's distinction between Skinnerian and Popperian creatures, where the former learn only through reinforcement from previous actions, and the latter can mentally simulate actions prior to engaging in them. Thanks to Gabe Dupre for suggesting this point to me.

it has wide applicability throughout the living world. Here I am mostly interested in describing features of the human that differentiate the human process from the kinds of D-sphexish, mechanistic, stimulus-response behaviour still often thought to apply to everything up to and perhaps including human action. Recalling my earlier identification of agent causation, the possibility of causal chains that originate in the agent, I have tried to sketch something of the kind of agent to whom such a capacity might plausibly be attributed. It is time to go into some more detail as to how this might work.

Let me first remind you that I am envisaging a world in which order, causal or otherwise, is an exceptional condition rather than an all-pervasive background. Persistent or continuant processes create much of the order that they exhibit. In some cases, what I have called agents, they may be the origin of causal chains, and these often serve to promote their persistence. If an organism is a process with capacities to respond to external contingencies in effective and even sometimes novel ways, then it is a source of causal processes in roughly the sense of agent causation.

This may, I suppose, still seem a bit mysterious. Organisms alleged to be the sources of causal chains still have causal histories and histories which, at least partially, explain the capacities they now have and their exercise of them at a particular point. How do we individuate some event within this forwards- and backwards-looking causal nexus as a point of origin of a new causal process? I hope that at least a sketch of an answer to these questions will emerge in the following sections of this chapter.

Goals

When I describe behaviour as effective, I do not mean, in the technical sense, 'adaptive', simply conducive to the familiar evolutionary ends of survival and reproduction; I mean whatever ends the organism may have. Frequently, these do involve the creation or maintenance of some kind of order conducive to the well-being of the organism. The self-evident fact, however, that humans have a great many aims, most of which have no connection to biological competition, is a central reason why sociobiology and Evolutionary Psychology have generally seemed hopelessly simplistic approaches to understanding the human.

So humans have many distinct goals, depending at least on their place in society and the division of labour. Farmers will have goals connected with food production; politicians, occasionally, with a well-run society; scientists and philosophers with an understanding of the world; and so on. Any of these people may have multiple other goals, from earning a golf handicap of ten to playing the accordion, from memorizing the Bible to improving the lives of domestic chickens. And here, I suggest, we find the highest level of agency, perhaps unique to humans in our world. It is by virtue of acting in ways conducive to the achievement of these multiple goals that humans are most distinctively sources of causal power, or free agents.

How does all this connect with the difficulties with free will outlined at the beginning of this chapter? Part of the traditional difficulty was with seeing humans as somehow exceptions to an otherwise seamless web of deterministic causality, a view that has rightly seemed metaphysically desperate. But I have proposed to reverse this. There is no seamless web of causal order, but a background of disorder out of which threads of partial order here and there emerge. Humans are exceptions to the general disorder and lawlessness of the universe; not unique exceptions certainly, but nonetheless, I suggest, we are the densest concentrations of causal power in the known universe. Surely there is room to better understand the way that causal order emerges from the surrounding disorder, but understanding the human condition is continuous with this project rather than an exception to it.

If this seems surprising, think for a moment of how extraordinarily *predictable* humans are. If I arrange to meet you for a cup of coffee at a particular time and place three weeks from now, I am fairly confident that you will be there, even if you have been all over the world in between. Think what a remarkable achievement this is. It would be even more remarkable in a deterministic world—think of all the calculations required—and only becomes seriously possible in a malleable world in which the predictable entity can constantly adjust its behaviour to emerging contingencies.

So can this view escape the apparently watertight dilemma of human action being either causally explicable, and thus not open to any alternative, or randomly generated and thus in no way attributable to the agent? Watertight, because any relaxation of the causal explicability of the action seems only to introduce an unwanted element of randomness.

I don't think this dilemma can be escaped entirely, but I think it can be rendered relatively innocuous. What it does show is that there is an important truth to compatibilism. Humans really cannot escape the causal order, whatever that is. If, as I argue, causal order is rare, but humans are replete with it, this does not, obviously, put humans outside that order. But if the causality that generates our action is our own capacity to impose order on our world, then I don't think we should be very concerned: the order that constrains us is largely of our own making.

But the recurrent dilemma still won't go quietly. Much of the concern here comes from questions of responsibility, credit, and blame. I perform a noble or vile act and perhaps deserve credit or blame. If I could not have done otherwise, is this desert justified? The problem is that even if I could have done otherwise it is far from obvious how this helps. Why did I do what I did, whether or not I could have refrained? The unhelpful dichotomy of cause and chance still confronts us.

We begin to make progress, I think, by distinguishing between kinds of motivation. To the extent that I do exactly and only just what I feel moved to do at the moment, following whatever urges I may have, I see no very interesting use for the distinction between free and unfree action. But processes can contribute to shaping themselves. A river may erode a bank thereby providing a quicker route to the ocean, for example. Humans, rather more interestingly, shape their path through the world with decisions on goals and policies. If I decide to go to the gym three times a week, perhaps to improve my physical fitness, I plan to constrain my choices. When I decide whether to stay in bed another hour, or stick to my plan of going to the gym, I make a real choice. The ability to do not what I am currently moved to do, stay in bed, but what I am committed to doing, go to the gym, seems to me the interesting exercise of freedom. As a process creating a path of order through a chaotic world, I am able to choose some part of the order I impose.

This idea connects free will clearly with the general theme of these chapters. The history of the universe, and the history of life as a vital phase in that history, is one of the emergence of order from chaos. But the stable, ordered processes that have emerged during this long history are not merely the consequences of some now complete process, but are also the causes of further emergence of order. The emergence of hydrogen and helium atoms, stars, heavier atoms, organic molecules, cells, multicellular organisms, and so on have all brought with them new capacities that have made possible new forms of organization. The human capacity to acquire understanding of the workings of the world, and fundamentally reorganize it if, sadly, now in directions threatening great harm to ourselves, is a new stage in this gradual emergence of causal powers.

Many distinctively human powers can be exercised by the person who simply does that which they feel most strongly inclined to do at the present moment. But most of what is uniquely human, the building of railway lines and skyscrapers, the discovery of the structure of matter, and so on, is possible only through commitment to override the desires of the moment. (And, incidentally, through massive cooperation.) This idea is clearly related to Kant's view that only in acting in accordance with moral duty am I free. But my view is much more tolerant than Kant as to what kinds of principle or goal can, as he puts it, provide a maxim for my action. And I offer no view on whether there is an objective duty to which I am subject. What my view shares with Kant is the perhaps paradoxical idea that it is *limiting* my possible action in accordance with some larger ambition for my life that provides freedom. An obvious advantage the view shares with Kant's more austere position is that it properly relates freedom of the will with the will, something that can be trained and cultivated, and something that can be subject to strength or weakness (akrasia). A strong will is what enables one to align one's behaviour with one's principles and goals rather than one's first order desires. It is by doing so that humans and perhaps some other animals impose new forms of order in the disordered world.

I said the recurrent question had been pushed back, but it does not go away. Where do my principles or my goals come from. The answer, I take it, is from a combination of upbringing, perhaps genetic luck, and experience. But I did not choose these. So can I claim any credit for either the admirable or contemptible principles that guide my actions, or the actions that these evoke? I think that with one important proviso this is where the defence of freedom must end. We admire strength of will and good character, but it is questionable whether we should give people credit for these in the way we do for actions and courses of action. Similarly we admire natural talents, whether athletic or intellectual, without properly giving people credit for them. The proviso is that in addition to endowments that are a matter of luck, people are to some important degree self-creating. Self-cultivation, as has been noted for millennia, is a vital aspect of human lives, and such things as strength of will are certainly to some degree products of this. But yet again, the ability to cultivate the self is itself something that must depend in large part on accidents of origin. How much of this is a variable within individual control is undoubtedly a crucial question in moral psychology, but not one I can attempt to pursue here beyond noting that it is surely something that belongs firmly within the remit of process philosophy.

Cooperation Again: Freedom as Social Construct

I want to conclude with one very important point about human agency. I have spoken of the various projects that can shape the second order desires that differentiate the free agent from the consistent follower of the desire of the moment. I have in mind a vast array of such projects from the pursuit of learning or spiritual insight, to building houses or roads, to curing illness and teaching the young. With regard to any of these it is crucial to recall the points I have stressed before on the cooperative nature of the human species. Any of these projects, even the most self-directed, depend on the contributions of innumerable fellow humans, whether in providing us the means of existence in food, housing, or clothing, the means of development into competent adult humans, through teaching, medical provision, and suchlike, or the material infrastructure of housing, transport, and much else. The possibility of self-cultivation, with which I ended the main discussion of human freedom, is itself something entirely dependent on all of these social inputs.

The point of all this is that freedom is, in the end, a social construct. I do not mean here, as so often and strangely seems to be taken as an interpretation of social construction, not real. Society is real, and construction is, generally at least, a way of making things real (Dupré 2004). I mean rather that the freedom we enjoy is not to be understood primarily as the expression of the essence of a unique individual, but as the realization by the individual of something the enabling conditions of which are overwhelmingly social. As I have mentioned at several points in these chapters, a key conclusion of the processual perspective on human life is that the emphasis on the isolated individual that has been such a central feature of so much Western thought for the last few centuries is seriously misguided. I do not mean that we should not cherish the uniqueness and unique contributions of the individual. But we should also remember how the ability to enjoy these things is entirely dependent on a functioning society. Indeed, I have contrasted the pursuit of grander goals with the submission to passing desire, but the ability to make this choice is always contingent on the background of passing wants—food, water, clothing, shelter—being generally provided. There is little freedom for the destitute; freedom is a luxury reserved for those with access to the benefits of a functioning society. Perhaps this should defuse some of our worries at the unanswered questions at the end of the discussion of individual free will. Certainly, it should reinforce the awareness that the grossly unequal distribution of resources in contemporary societies is not only wholly unjustified by any difference in individual contributions, but also damaging to the prospects for rewarding lives for most of us.

Afterword

This book has moved, no doubt too quickly, from the global to the very local; from the most general questions about what the universe is made of, what kinds of entities it contains, to questions about the very particular and peculiar nature of one species on a planet on the outer fringes of a not particularly exceptional galaxy, *Homo sapiens*: perhaps an exceptional species, but certainly of exceptional interest to us, its members. Though it hardly needs saying that much in between, and even much at either end of this spectrum, has been treated with complete superficiality, I do not mean to apologize for the broad scope. A central message of the book is that there is a way of looking at all the diverse phenomena in our universe that provides a coherent picture, namely the process metaphysics that I have been aiming to motivate. Though I have spent much of my professional career emphasizing the diversity of phenomena, this should not prevent us from seeing that at a sufficiently abstract level, the level of (naturalistic) metaphysics, there is a coherent picture that ties them together. The diversity of phenomena explains the inevitable plurality of our science, but this plurality should not lead us into denying that there is one coherent world (Dupré 2024). This world, I propose, is one in which partially ordered and stable processes, processes with novel causal powers, have gradually emerged from a chaotic background.

I then turned to life and argued in much greater detail that this picture of the emergence of partially stable processes was the correct way to understand this remarkable phenomenon. I first argued this for the case of organisms, and then for the slightly less familiar case of lineages, the long-lasting process in which evolution occurs. One important point of this exercise was to question assumptions about these entities that derive not from empirical data but from bad metaphysics. So, for example, the idea that an individual or a species has an essential property, the assumption that there are sharp boundaries, both spatial and temporal, to an organism, that organisms are largely autonomous, and so on, are all ideas that derive from the metaphysics of substance, but actually contradict the deliverances of empirical science. Indirectly, this provides fuel for an argument for the ultimate authority of a metaphysics grounded in empirical, often scientific, knowledge, a naturalistic metaphysics, over the a priori metaphysics still widely advocated.

Having established, to the best of my ability, the conclusion that life was a realm of mutually intertwined and stabilizing processes, I explored the implication that we ourselves, humans, must also be processes. This exploration was divided into three parts. First, I looked at the boundaries of the individual human, both spatial but even more temporal, and the vagueness of these in a world of process. I suggested that these provided fresh insight into traditional questions about personal identity. Second, I looked at the classification of humans and the scepticism about such classifications engendered by a process perspective. And third, I explored the implications of a process ontology for, generally, the understanding of the agency of living beings and, finally, the ancient question of freedom of the will.

Throughout these investigations I have tried to stick to evidence-based argument. As a philosopher, I do not mean to downplay the argument part of this methodology; but I also want to insist that where contingent propositions about the world are concerned, we should defer to the evidence and, with due caution, this generally means deferring to science. But as my reflections come to an end, I would like to take a rather different turn, and admit that one reason I am drawn to defend a process ontology is that I find that it portrays a much more appealing world than its more traditional alternative.

While I have tried to convince you of the inescapability of process philosophy by argument, of course there are also less purely rational reasons why we hold particular views. And I confess that I like the process world. Why? Billiard ball science and metaphysics—as I refer to the substance view of things with deliberate abuse—describes a world of necessity, a world governed by law. Science, in such a world, tells us how things are and have to be. Sometimes as in cases such as neoclassical economics or Evolutionary Psychology, the necessity is oppressive. We cannot make a fairer society, we are told, because we are prevented by the facts of human nature or the iron laws of economics. Process science is a science of possibility. Processes evolve, and their evolution is affected by many factors. There are very many things in the present world that are not as we would wish them to be, and process science can, perhaps, tell us how we might help them to get better. It may not be easy, but neither is it ruled out from the start by philosophical argument. This is surely a more attractive and encouraging view of the world. How fortunate that it is also true!

Acknowledgements

A theme of this book is that human activity is vastly more cooperative than is usually assumed by the tradition that sees us as independent and autonomous substances, and the book itself is an excellent illustration of this thesis. Far more people than I can thank have contributed to making it possible during the several decades over which the ideas in the book have developed. I shall, at least, mention some of the more recent and more important contributions.

The natural starting point to mark the beginning of this long process is the award of a centre grant from the UK Economic and Social Research Council (ESRC), which established Egenis, The ESRC Centre for Genomics in Society. The ESRC Centre both funded my own work for ten years (2002–2012) and allowed me to assemble a wonderful group of colleagues and a perfect working environment that has survived their generosity to the present day. After the ten years of ESRC support, Egenis reorganized itself as the Centre for the Study of Life Sciences, with a broader concern with the whole range of biological sciences. My appointment for three months in 2006 to a Spinoza Professorship at the University of Amsterdam allowed me a break from the day-to-day running of Egenis. As mentioned in the Introduction, it was in Amsterdam that my thoughts on process first came into clear focus, a focus that has continued to inform my work up to and including this book.

My work on process was subsequently helped enormously by an Advanced Investigator Award from the European Research Council (ERC award # 324186), which enabled me to spend five years from 2013 to 2018 fully devoted to the project, 'A Process Ontology for Biology'. Apart from continuing to give me enviable time to pursue my own research, this enabled me to collaborate over most of that time with three outstanding postdocs and thereby advance the research project much further than I could have on my own.

Daniel Nicholson provided, among other things, historical expertise on the important period in the early-to-mid-twentieth century when a number of biologists, most notably C. H. Waddington, were working on biological theory inspired by the leading process philosopher of the twentieth century A. N. Whitehead. I owe him several excellent quotes from this period. Dan has also done some of the best recent work critical of the new mechanism

(Nicholson 2012, 2013, 2014), drawing on the organicism developed by those earlier scientists and philosophers, and has articulated in detail an alternative processual account of the organism (Nicholson 2018). He and I co-edited the major output of the ERC project, the edited volume *Everything Flows* (2018). The introductory essay we co-authored, 'A Manifesto for a Processual Philosophy of Biology', remains the best (reasonably) concise summary of the project.

Anne Sophie Meincke provided us with an expert knowledge of the mainstream metaphysics with which we were directly or indirectly engaging or disputing. She developed an important and compelling argument that a substance metaphysics and the standard accounts of the persistence over time of a thing cannot give a coherent account of change (Meincke 2019). She has published a number of papers on personal identity and biological autonomy that are highly complementary to chapters 4 and 6 of this book (e.g. Meincke 2018, 2021). We also co-edited a volume of papers on biological identity (Meincke and Dupré 2021) by both philosophers of biology and more traditional metaphysicians, exploring common ground and important differences between points of view from these disciplines.

Perhaps the most challenging part of biology for the process philosopher is biochemistry, the area from which the new mechanists, perhaps the most important competing school in the metaphysics of biology, draw many of their favourite examples. Stephan Guttinger, bringing with him an expert background in biochemistry, has provided powerful arguments for the necessity of a processual understanding even of this most mechanistic-looking area of science (Guttinger 2018, 2021). I have also worked with him on the topic of viruses, which I think are one of the best examples of a biological process sometimes mistaken for a thing (Dupré and Guttinger 2016). I am delighted to be able to say that Stephan has returned to Exeter as a colleague.

In the Winter of 2023, I had the great privilege of being invited to take up a Donald Gordon Fellowship at the Stellenbosch Institute for Advanced Studies (STIAS) in South Africa, where I drafted what became the Gifford Lectures at Edinburgh later that year, and subsequently this book. STIAS is everything that a Research Centre should be. It is a wonderful community of scholars and hub of interdisciplinary conversation. The location is stunning, the resources superb, and the staff as helpful, friendly, and competent as one could hope. Even the food is delicious. I must particularly thank the Director, Edward Kirumira, the Programme Manager, Christoph Pauw, and Nel-Mari Loock, whose title (Coordinator: Information management, Fellows' and IT support and office arrangement), gives the correct impression that she seems to do everything. I would also like to thank all the staff who contributed to the wonderfully warm and supportive atmosphere of STIAS. There could not be a better place to undertake an extensive piece of academic work.

Which leads me naturally to expressing my gratitude to the University of Edinburgh and the Gifford Lectures Committee. It is a great privilege to be invited to contribute to this distinguished lecture series and humbling to see the list of my predecessors in this series, and I am most grateful to the Committee for so honouring me. My greatest personal debt is to Stewart (Jay) Brown, who as well as extensive correspondence ahead of my lectures, hosted the entire two weeks of my visit with extreme graciousness, helpfulness, and good company. I am also most grateful to Louise Trotter for her tireless efficiency in dealing with countless details of the visit.

I must also thank: Sarah Prescott, Steven Yearley, Sir Peter Mathieson, Lesley McAra, Mark Harris, and Jay Brown for chairing the lectures; Victoria Turner for organizing the blog that ran parallel to the lectures, as well as all the authors of the many excellent contributions to the blog; and the audiences that asked me many interesting and challenging questions. It was, in all respects, a thoroughly enjoyable and stimulating occasion. The chapters in this book are closely based on the lectures, without which the book would not exist.

There are, as I already mentioned, many more people who contributed to these lectures and the subsequent book than I can thank here. I have discussed the topics in various venues and publications and have received comments and questions that I'm sure have influenced my thinking even as I may have forgotten who contributed. So thanks to all the many colleagues who have engaged in large or small ways with my philosophical work. A particular debt should be mentioned to the scientists who have taken an interest in philosophical work on biology in general, and my own in particular. I think especially of Ford Doolittle, Scott Gilbert, and James Wakefield.

Most of all, I would like to thank my colleagues in Egenis, The Centre for the Study of Life Sciences, who have provided an invariably stimulating and enjoyable intellectual environment. Very special thanks to Sabina Leonelli, who took over the Directorship of the Centre two years ago, and brought it to a new level of intellectual activity. My intellectual collaboration with her also led me to a first attempt to engage properly with the biggest lacuna in my work on process, process epistemology, or the methodological implications of a process metaphysics (Dupré and Leonelli 2022), with reference to problems with the science around COVID-19 that arose, we argued, from a misguided thing ontology. Sadly, she left us for new challenges in 2024, and

I wish her a wonderful new life in Munich. Special thanks also to Celso Neto for applying his expertise on the philosophy of race to a draft of chapter 5, to my long-time collaborator, artist Gemma Anderson, who provided the excellent illustrations for chapter 2 and to George Newman for his excellent work on the index.

Very numerous current colleagues in Egenis have also contributed greatly to my intellectual life over the last twenty years. Special thanks to the cofounders of the Centre, Barry Barnes and Steve Hughes, without whom, of course, none of this could have happened. And many thanks to numerous past and present colleagues, including Gemma Anderson, Tyler Brunet, Jane Calvert, Adrian Currie, Hannah Farrimond, Christine Hauskeller, Susan Kelly, Sabina Leonelli, Staffan Müller-Wille, Maureen O'Malley, Celso Neto, Ginny Russell, Paula Saukko, Adam Toon, and Özlem Yilmaz. Thanks to several PhD students who have worked with me on a wide variety of topics, always broadening my knowledge and interests: Ann-Sophie Barwich, David Batty, Thomas Bonnin, Jonathan Davies, Javier Suarez Diaz, Flavia Fabris Jack Griffiths, Elis Jones, Čaglar Karaca, Steffan Llewellyn, Pierre-Olivier Méthot, Mila Petrova, Thibault Racovski, Ric Sims, and Kai Wang. My apologies to those I have forgotten to include. My never great and now declining memory make perfection in this pleasant task impossible. Finally, two anonymous referees for Oxford University Press provided very generous reviews with suggestions that I have done my best to accommodate.

The experience of the social nature of the human species starts, for many of us, with the family. I have been more than blessed in this respect, and my debt to my wife and sons, Regenia, Gabe, and Julian, is immeasurable. Regenia and Gabe are brilliant academics, and the book has benefited immeasurably from their detailed readings of a draft of the lectures. Julian is a civil servant; his critical conversation and his effortless charm have contributed equally to the environment that has enabled me to write this book. Finally, just as the book was going to press, I was blessed with my first grandchild, Jackson. In difficult times for so many, it is wonderful to be reminded that there are always new beginnings with new possibilities and new hope. I dedicate this book to Jackson.

References

Anscombe, G. E. M. (1971). *Causality and Determination: An Inaugural Lecture.* Cambridge University Press Archive.

Austin, C. J. (2020). Organisms, Activity, and Being: On the Substance of Process Ontology. *European Journal for Philosophy of Science* 10: 13.

Bapteste, E. and Anderson, G. (2018). Intersecting Processes are Necessary Explanantia for Evolutionary Biology, but Challenge Retrodiction. In Nicholson and Dupré (2018), pp. 283–299.

Barkow, J. H., Cosmides, L., and Tooby, J. (eds). (1995). *The Adapted Mind: Evolutionary Psychology and the Generation of Culture.* New York: Oxford University Press.

Beiler, K. J., Durall, D. M., Simard, S. W., Maxwell, S. A., and Kretzer, A. M. (2010). Architecture of the Wood-Wide Web: Rhizopogon Spp. Genets Link Multiple Douglas-fir Cohorts. *New Phytologist* 185: 543–553.

Bobay, L.-M. and Ochman, H. (2017). Biological Species are Universal across Life's Domains. *Genome Biology and Evolution* 49: 1–501.

Bostrom, N. (2003). Are We Living in a Computer Simulation? *The Philosophical Quarterly* 53: 243–255.

Boyle, Robert. (1666). *The Origine of Formes and Qualities (According to the Corpuscular Philosophy) Illustrated by Considerations and Experiments.* Oxford: H. Hall printer to the University.

Brockmann, H. J. (1985). Provisioning Behavior of the Great Golden Digger Wasp, *Sphex ichneumoneus* (L. Sphecidae). *Journal of the Kansas Entomological Society* 58: 631–655.

Brunet, T. and Doolittle, F. (2018). The Generality of Constructive Neutral Evolution. *Biology and Philosophy* 33: 1–25.

Buller, D. J. (2006). *Adapting Minds: Evolutionary Psychology and the Persistent Quest for Human Nature.* Cambridge, MA: MIT press.

Calvo, P., Gagliano M., Souza, G. M., and Trewavas, A. (2020). Plants are Intelligent, Here's How. *Annals of Botany* 125: 11–28.

Chen, E. K. (2023). Does Quantum Theory Imply the Entire Universe is Preordained? *Nature* 624(7992): 513–515.

Clark, R. D. and Hatfield, E. (1989). Gender Differences in Receptivity to Sexual Offers. *Journal of Psychology & Human Sexuality* 2: 39–55.

Clarke, E. (2016). Levels of Selection in Biofilms: Multispecies Biofilms are not Evolutionary Individuals. *Biology and Philosophy* 31: 191–212.

Clarke, R. (1993). Toward a Credible Agent-Causal Account of Free Will. *Noûs* 27: 191–203.

Clutton-Brock, T. (2009). Cooperation Between Non-Kin in Animal Societies. *Nature* 462: 51–57.

Darwin, C. (1881). *The Formation of Vegetable Mould Through the Action of Worms with Some Observations on Their Habits.* London: John Murray.

Davis, J. N. and Daly, M. (1997). Evolutionary Theory and the Human Family. *The Quarterly Review of Biology* 72: 407–435.

Dawkins, R. (1976). *The Selfish Gene.* Oxford: Oxford University Press.

Dawkins, R. (2006). *Climbing Mount Improbable*. London: Penguin.

De Grey, A. and Rae, M. (2007). *Ending Aging: The Rejuvenation Breakthroughs that Could Reverse Human Aging in Our Lifetime*. New York: St. Martin's Press.

de La Mettrie, J. O. (1994 [1747]). *Man a Machine; And, Man a Plant*. Indianapolis: Hackett Publishing.

Dennett, D. (1973). Mechanism and Responsibility. In *Essays on Freedom of Action*, Honderich, T. (ed.). London: Routledge.

Donlan, R. M. (2002). Biofilms: Microbial Life on Surfaces. *Emerging Infectious Diseases* 8: 881.

Doolittle, W. F. (1999). Phylogenetic Classification and the Universal Tree. *Science* 284(5423): 2124–2128.

Dupré, J. (1993). *The Disorder of Things: Metaphysical Foundations of the Disunity of Science*. Cambridge, MA: Harvard University Press.

Dupré, J. (2001). *Human Nature and the Limits of Science*. Oxford: Oxford University Press.

Dupré, J. (2004). What's the Fuss about Social Constructivism? *Episteme* 1: 73–85. (Reprinted in Dupré 2012.)

Dupré, J. (2008). *The Constituents of Life*. Amsterdam: Koninklijke Van Gorcum. (Reprinted in Dupré 2012.)

Dupré, J. (2011). Emerging Sciences and New Conceptions of Disease; Or, Beyond the Monogenomic Differentiated Cell Lineage. *European Journal for Philosophy of Science* 1: 119–131. (Reprinted in Dupré 2012.)

Dupré, J. (2012). *Processes of Life: Essays in the Philosophy of Biology*. Oxford: Oxford University Press.

Dupré, J. (2014). Animalism and the Persistence of Human Organisms. *The Southern Journal of Philosophy* 52: 6–23.

Dupré, J. (2021a). *The Metaphysics of Biology*. Cambridge: Cambridge University Press.

Dupré, J. (2021b). Causally Powerful Processes. *Synthese* 199: 10667–10683.

Dupré, J. (2024). The Disunity of Science and the Unity of the World. Presidential Address, PSA 2022. *Philosophy of Science* 91: 1043–1057.

Dupré, J. and Guttinger, S. (2016). Viruses as Living Processes. *Studies in History and Philosophy of Science Part C: Studies in History and Philosophy of Biological and Biomedical Sciences* 59: 109–116.

Dupré, J. and Leonelli, S. (2022). Process Epistemology in the COVID-19 Era: Rethinking the Research Process to Avoid Dangerous Forms of Reification. *European Journal for Philosophy of Science* 12: 20.

Dupré, J. and O'Malley, M. A. (2009). Varieties of Living Things: Life at the Intersection of Lineage and Metabolism. *Philosophy and Theory in Biology* 1. <http://dx.doi.org/10.3998/ptb.6959004.0001.003>

Emlen, S. T. (1995). An Evolutionary Theory of the Family. *Proceedings of the National Academy of Sciences* 92(18): 8092–8099.

Ereshefsky, M. and Pedroso, M. (2013). Biological Individuality: The Case of Biofilms. *Biology and Philosophy* 28: 331–349.

Fabre, J.-H. (2001 [1915]). *The Hunting Wasps*. New York: Dodd, Mead, and Company.

Fitz-James, M. H. and Cavalli, G. (2022). Molecular Mechanisms of Transgenerational Epigenetic Inheritance. *Nature Reviews Genetics* 23: 325–341.

Flemming, H.-C. and Wingender, J. (2010). The Biofilm Matrix. *Nature Reviews Microbiology* 8: 623–633.

Foucault, M. (1997). *Ethics: Subjectivity and Truth*. Rabinow, P. (ed.). New York: The New Press.

Gagnier, R. (2000). The Law of Progress and the Ironies of Individualism in the Nineteenth Century. *New Literary History* 31: 315–336.

Ghiselin, M. T. (1974). A Radical Solution to the Species Problem. *Systematic Biology* 23: 536–544.

Gilbert, S. F. and Epel, D. (2015). *Ecological Developmental Biology: The Environmental Regulation of Development, Health and Evolution* (2nd edition). Sunderland, MA: Sinauer.

Glennan, S. S. (1996). Mechanisms and the Nature of Causation. *Erkenntnis* 44: 49–71.

Godfrey-Smith, P. (2009). *Darwinian Populations and Natural Selection*. New York: Oxford University Press.

Griffiths, P. and Stotz, K. (2018). Developmental Systems Theory as a Process Theory. In Nicholson and Dupré (2018), pp. 225–245.

Griffiths, P. E. and Gray, R. D. (1994). Developmental Systems and Evolutionary Explanation. *The Journal of Philosophy* 91: 277–304.

Guttinger, S. (2018). A Process Ontology for Macromolecular Biology. In Dupré and Nicholson (2018), pp. 303–320.

Guttinger, S. (2021). Process and Practice: Understanding the Nature of Molecules. *HYLE: International Journal for Philosophy of Chemistry* 27: 47–66.

Hacking I. (1995). The Looping Effects of Human Kinds. In *Causal Cognition*, Sperber, D. and Premack, A. J. (eds). Oxford: Oxford University Press, pp. 351–383.

Hannon, E. and Lewens, T. (eds). (2018). *Why We Disagree about Human Nature*. Oxford: Oxford University Press.

Hehemann, J.-H., Correc, G., Barbeyron, T., Helbert, W., Czjzek, M., and Michel, G. (2010). Transfer of Carbohydrate-Active Enzymes from Marine Bacteria to Japanese Gut Microbiota. *Nature* 464: 908–912.

Hodgson, G. M. (1996). *Economics and Evolution: Bringing Life back into Economics*. Ann Arbor: University of Michigan Press.

Hugenholtz, P., Chuvochina, M., Oren, A., Parks, D. H., and Soo, R. M. (2021). Prokaryotic Taxonomy and Nomenclature in the Age of Big Sequence Data. *The ISME Journal* 15: 1879–1892.

Hull, D. L. (1967). The Metaphysics of Evolution. *The British Journal for the History of Science* 3: 309–337.

Hull, D. L. (1976). Are Species Really Individuals? *Systematic Zoology* 25: 174–191.

Hull, D. L. (1999). On the Plurality of Species: Questioning the Party Line. In *Species: New Interdisciplinary Essays*, Wilson, R. A. (ed.). Cambridge, MA: MIT Press, pp. 3–20.

Hume, D. (1993 [1772]). *An Enquiry Concerning Human Understanding*. Indianapolis: Hackett.

Ingalhalikar, M., Smith, A., Parker, D., Satterthwaite, T. D., Elliott, M. A., Ruparel, K., Hakonarson, H., Gur, R. E., Gur, R. C., and Verma, R. (2014). Sex Differences in the Structural Connectome of the Human Brain. *Proceedings of the National Academy of Sciences* 111: 823–828.

Jablonka, E. and Lamb, M. J. (2014). *Evolution in Four Dimensions: Genetic, Epigenetic, Behavioral, and Symbolic Variation in the History of Life* (revised edition). Cambridge, MA: MIT press.

Jablonski, N. G. and Chaplin, G. (2000). The Evolution of Human Skin Coloration. *Journal of Human Evolution* 39: 57–106.

Jablonski, N. G. and Chaplin, G. (2017). The Colours of Humanity: The Evolution of Pigmentation in the Human Lineage. *Philosophical Transactions of the Royal Society B: Biological Sciences* 372: 20160349.

Keijzer, F. (2013). The Sphex Story: How the Cognitive Sciences Kept Repeating an Old and Questionable Anecdote. *Philosophical Psychology* 26: 502–519.

Kingma, E. (2018). Lady Parts: The Metaphysics of Pregnancy. *Royal Institute of Philosophy Supplements* 82: 165–187.

Kingma, E. (2019). Were You a Part of Your Mother? *Mind* 128: 609–646.

Lancy, D. F. (2014). *The Anthropology of Childhood: Cherubs, Chattel, Changelings.* Cambridge: Cambridge University Press.

Lax, E. (1975). *Woody Allen and His Comedy.* New York: Manor Books.

Lewontin, R. C. (1972). The Apportionment of Human Diversity. *Evolutionary Biology* 6: 381–398.

Liang, G. and Bushman, F. D. (2021). The Human Virome: Assembly, Composition and Host Interactions. *Nature Reviews Microbiology* 19: 514–527.

Liu, C. (2015). *The Three-Body Problem.* London: Bloomsbury Publishing.

Locke, J. (1975 [1689]). *An Essay Concerning Human Understanding.* Oxford: Clarendon Press.

Machamer, P., Darden, L., and Craver, C. F. (2000). Thinking about Mechanisms. *Philosophy of Science* 67: 1–25.

Mallet, J. (2005). Hybridization as an Invasion of the Genome. *Trends in Ecology & Evolution* 20: 229–237.

Mandelbaum, E. (2022). Everything and More: The Prospects of Whole Brain Emulation. *The Journal of Philosophy* 119: 444–459.

Margulis, L. (1981). *Symbiosis in Cell Evolution: Life and its Environment on the Early Earth.* San Francisco: W. H. Freeman and Co.

Margulis, L. and Fester, R. (1991). *Symbiosis as a Source of Evolutionary Innovation.* Cambridge, MA: MIT Press.

Mayr, E. (1942). *Systematics and the Origin of Species.* New York: Columbia University Press.

Meincke, A. S. (2018). Persons as Biological Processes: A Bio-processual Way out of the Personal Identity Dilemma. In Nicholson and Dupré (2018), pp. 357–378.

Meincke, A. S. (2019). The Disappearance of Change: Towards a Process Account of Persistence. *International Journal of Philosophical Studies* 27(1): 12–30.

Meincke, A. S. (2021). Processual Animalism: Towards a Scientifically Informed Theory of Personal Identity. In Meincke and Dupré (2021), pp. 251–278.

Meincke, A. S. (2022). One or Two? A Process View of Pregnancy. *Philosophical Studies* 179: 1495–1521.

Meincke, A. S. (2023). The Metaphysics of Development and Evolution. From Thing Ontology to Process Ontology. *Human Development* 67(5–6): 233–256.

Meincke, A. S. and Dupré, J. (eds) (2021). *Biological Identity: Perspectives from Metaphysics and the Philosophy of Biology.* Abingdon and New York: Routledge.

Mesoudi, A., Whiten, A., and Laland, K. N. (2006). Towards a Unified Science of Cultural Evolution. *Behavioral and Brain Sciences* 29: 329–347.

Méthot, P.-O. and Alizon, S. (2014). What is a Pathogen? Toward a Process View of Host-Parasite Interactions. *Virulence* 5: 775–785.

Miller, M. B. and Bassler, B. L. (2001). Quorum Sensing in Bacteria. *Annual Reviews in Microbiology* 55: 165–199.

Morgan, W. (2022). Are Organisms Substances or Processes? *Australasian Journal of Philosophy* 100: 605–619.

Müller, G. B. (2007). Evo–Devo: Extending the Evolutionary Synthesis. *Nature Reviews Genetics* 8: 943–949.

Mumford, S. and Anjum, R. L. (2011). *Getting Causes from Powers*. New York: Oxford University Press.

Nadell, C. D., Drescher, K., and Foster, K. R. (2016). Spatial Structure, Cooperation and Competition in Biofilms. *Nature Reviews Microbiology* 14: 589–600.

Nicholson, D. J. (2012). The Concept of Mechanism in Biology. *Studies in History and Philosophy of Science Part C: Studies in History and Philosophy of Biological and Biomedical Sciences* 43(1): 152–163.

Nicholson, D. J. (2013). Organisms≠ Machines. *Studies in History and Philosophy of Science Part C: Studies in History and Philosophy of Biological and Biomedical Sciences,* 44(4): 669–678.

Nicholson, D. J. (2014). The Machine Conception of the Organism in Development and Evolution: A Critical Analysis. *Studies in History and Philosophy of Science Part C: Studies in History and Philosophy of Biological and Biomedical Sciences* 48: 162–174.

Nicholson, D. J. (2018). Reconceptualizing the Organism: From Complex Machine to Floere Stream. In Nicholson and Dupré (2018), pp. 139–166.

Nicholson, D. J. and Dupré, J. (eds) (2018). *Everything Flows: Towards a Processual Philosophy of Biology*. Oxford: Oxford University Press.

O'Connor, T. and Franklin, C. (2022). Free Will. *The Stanford Encyclopedia of Philosophy* (Winter 2022 edition), Zalta, E. N. and Nodelman, U. (eds), <https://plato.stanford.edu/archives/win2022/entries/freewill/>.

Oderberg, D. S. (1997). Modal Properties, Moral Status, and Identity. *Philosophy and Public Affairs* 26: 259–298.

Odling-Smee, F. J., Laland, K. N., and Feldman, M. W. (2003). *Niche Construction: The Neglected Process in Evolution*. Princeton: Princeton University Press.

Olson, E. T. (1997). *The Human Animal: Personal Identity Without Psychology*. Oxford: Oxford University Press.

Oyama, S. (1985). *The Ontogeny of Information: Developmental Systems and Evolution*. Cambridge: Cambridge University Press.

Parfit, D. (1984). *Reasons and Persons*. Oxford: Oxford University Press.

Paterson, H. E. H. (1985). The Recognition Concept of Species. In *Species and Speciation*, Vrba, E. S. (ed.). Pretoria: Transvaal Museum (Transvaal Museum Monograph, 4), pp. 21–29.

Pinker, S. (2003). *The Blank Slate: The Modern Denial of Human Nature*. New York: Viking.

Pradeu, T. (2011). *The Limits of the Self: Immunology and Biological Identity*. Oxford: Oxford University Press.

Pradeu, T. (2019). Immunology and Individuality. *Elife* 8: e47384.

Reiss, J. (2009). *Not by Design: Retiring Darwin's Watchmaker*. Berkeley: University of California Press.

Robert, J. S., Hall, B. K., and Olson, W. M. (2001). Bridging the Gap Between Developmental Systems Theory and Evolutionary Developmental Biology. *BioEssays* 23: 954–962.

Robinson, H. (2021). Substance. *The Stanford Encyclopedia of Philosophy* (Fall 2021 edition), Zalta, E. N. (ed.), <https://plato.stanford.edu/archives/fall2021/entries/substance/>.

Roossinck, M. J. (2011). The Good Viruses: Viral Mutualistic Symbioses. *Nature Reviews Microbiology* 9: 99–108.

Rosenberg, N. A., Pritchard, J. K., Weber, J. L., Cann, H. M., Kidd, K. K., Zhivotovsky, L. A., and Feldman, M. W. (2002). Genetic Structure of Human Populations. *Science* 298: 2381–2385.

Sandberg, A. and Bostrom, N. (2008). *Whole Brain Emulation: A Roadmap*. Technical Report #2008-3, Future of Humanity Institute, Oxford University.

Sapp, J. (1994). *Evolution by Association: A History of Symbiosis.* New York: Oxford University Press.

Sellars, W. S. (1962). Philosophy and the Scientific Image of Man. In *Frontiers of Science and Philosophy*, Colodny, R. G. (ed.). Pittsburgh: University of Pittsburgh Press, pp. 35–78.

Shkoporov, A. N., Clooney, A. G., Sutton, T. D. S., Ryan, F. G., Daly, K. M., Nolan, J. A., McDonnell, S. A., et al. (2019). The Human Gut Virome is Highly Diverse, Stable, and Individual Specific. *Cell, Host and Microbe* 26: 527–541.

Simons, P. (2018). Processes and Precipitates. In Nicholson and Dupré (2018), pp. 49–60.

Simpson, G. G. (1961). *Principles of Animal Taxonomy.* New York: Columbia University Press.

Singh, D. (1993). Adaptive Significance of Female Physical Attractiveness: Role of Waist-to-Hip Ratio. *Journal of Personality and Social Psychology* 65: 293.

Skrzypek, J. W. (2023). Trust the Process? Hyloenergeism and Biological Processualism. *Ratio* 36(4): 334–346.

Smith, A. (1982 [1776]). *The Wealth of Nations.* London: Penguin.

Steward, H. (2020). Substances, Agents and Processes. *Philosophy* 95: 41–61.

Stout, R. (2016). The Category of Occurrent Continuants. *Mind* 125: 41–62.

Tooby, J. and Cosmides, L. (2005). Conceptual Foundations of Evolutionary Psychology. In *The Handbook of Evolutionary Psychology*, Buss, D. M. (ed.). Hoboken, NJ: John Wiley and Sons, pp. 5–67.

Van Valen, L. (1976). Ecological Species, Multispecies, and Oaks. *Taxon* 25: 233–239.

Waddington, C. H. (1957). *The Strategy of the Genes: A Discussion of Some Aspects of Theoretical Biology.* London: George Allen and Unwin.

Waring, M. (1988). *If Women Counted: A New Feminist Economics.* San Francisco: Harper & Row.

Warnock, M. (1984). *Report of the Committee of Inquiry into Human Fertilisation and Embryology.* Vol. 9314. London: HM Stationery Office.

West-Eberhard, M. J. (2003). *Developmental Plasticity and Evolution.* Oxford: Oxford University Press.

Wiggins, D. (2016). Activity, Process, Continuant, Substance, Organism. *Philosophy* 91: 269–280. Reprinted in Meincke and Dupré (2021).

Wilkins, J. S., Zachos, F. E., and Pavlinov, I. Ya. (eds). (2022). *Species Problems and Beyond: Contemporary Issues in Philosophy and Practice.* Boca Raton, FL: CRC Press.

Williamson, T. (2013). *Modal Logic as Metaphysics.* Oxford: Oxford University Press.

Young, T. P., Stubblefield, C. H., and Isbell, L. A. (1996). Ants on Swollen-Thorn Acacias: Species Coexistence in a Simple System. *Oecologia* 109: 98–107.

Yu, Y., Wan, Z., Wang, J.-H., Yang, X., and Zhang, C. (2022). Review of Human Pegivirus: Prevalence, Transmission, Pathogenesis, and Clinical Implication. *Virulence* 13: 323–340.

Zilber-Rosenberg, I. and Rosenberg, E. (2008). Role of Microorganisms in the Evolution of Animals and Plants: The Hologenome Theory of Evolution. *FEMS Microbiology Reviews* 32: 723–735.

Index

The manufacturer's authorised representative in the EU for product safety is Oxford University Press España S.A. of El Parque Empresarial San Fernando de Henares, Avenida de Castilla, 2 - 28830 Madrid (www.oup.es/en or product.safety@oup.com). OUP España S.A. also acts as importer into Spain of products made by the manufacturer.

Printed and bound by CPI Group (UK) Ltd, Croydon, CR0 4YY

10/06/2026

02133964-0001